HOTHOUSE EARTH

HOTHOUSE EARTH

Howard A. Wilcox

A Frank E. Taylor Book

PRAEGER PUBLISHERS

New York

Published in the United States of America in 1975
by Praeger Publishers, Inc.
111 Fourth Avenue, New York, N.Y. 10003

Library of Congress Cataloging in Publication Data

Wilcox, Howard A
  Hothouse earth.

  "A Frank E. Taylor book."
  Includes bibliographical references
  1. Thermal pollution of rivers, lakes, etc.
2. Heat budget (Geophysics) 3. Waste heat.
I. Title.
TD427.H4W54   363.6   74-15689
ISBN 0-275-52910-X

Printed in the United States of America

TO MY FAMILY, TO MY DESCENDANTS, AND TO ALL OUR DESCENDANTS,
IN THE HOPE THAT THEY (AND THROUGH THEM
ALL VALUABLE EARTHLY FORMS OF LIFE)
MAY CONTINUE THE GAME OF EVOLUTION
ON THIS PLANET AND IN THIS SOLAR SYSTEM

CONTENTS

ACKNOWLEDGMENTS

So many people have contributed importantly in one way or another to this book that, to my great regret, I cannot cite them all within a reasonably brief list.

Among climatologists, the persons who come mainly and immediately to my mind are Drs. J. Murray Mitchell, Jr., W. W. Kellogg, W. D. Sellers, Lester Machta, J. O. Fletcher, and M. I. Budyko. My understanding of physics in general, and of the prospects and problems of nuclear power in particular, stems mainly from my father's teaching plus that of my former professors and colleagues Drs. E. Fermi, S. K. Allison, E. Teller, and C. N. Yang. My concepts concerning the possibilities and realities of ocean farming were sharpened mainly by Dr. Wheeler North. Insight into the social dynamics of societies, governments, and industries has come to me mainly from my mother and from A. V. Butterworth.

Valuable criticism and encouragement have been contributed by such friends as Drs. W. F. and M. S. Cartwright, Dr. and Mrs. W. B. LaBerge, Dr. and Mrs. W. B. McLean, Drs. Richard Bogan, D. S. Potter, Howard Blood, and Carl Barus; also by Captain Gautier, Douglas Wilcox (no relation), Bruce and Brian Wilcox (my sons), Carol Wilcox (my daughter), Les Scher, Tom Leese, Ron Sweig, Ron Schneider, Drs. Norman Heidelbaugh, John Craven, Robert Cohen, Dr. and Mrs. Otto Heinz, Drs. R. L. Meier, Arthur Morgan, Richard Greeley, and Bill Hamilton, Mary Heathcote, Larry Brundall, Mr. and Mrs. John Werth, Dr. John Bardach, Mr. and Mrs. Peter Schauffler, Don Witcher, and Don Friedman.

My deep thanks go to Drs. Bill Aron, Ab Flowers, Richard

ACKNOWLEDGMENTS

Lehman, and J. Murray Mitchell, Jr., for especially valuable encouragement, support, and guidance.

This book would never have been realized without the steady vision, faith, and drive of Frank E. Taylor. Likewise, the support and assistance of my wife, Evelyn, have been as vital to this as to all my other enterprises.

Whatever readability and literary merits the book possesses are due mainly to my collaborator, John J. Fried. Any defects or shortcomings must, of course, be blamed solely on the author.

HOWARD A. WILCOX

San Diego
May, 1975

HOTHOUSE EARTH

1
INTRODUCTION

As the twentieth century staggers to its end, we are beginning to see that the next five to ten decades will confront humanity with three major challenges. First, we will be called upon to exercise great restraint in the way we use the vast natural storehouses of energy that are available to us, or else watch as our zealous consumption brings on a global heat disaster that could write an end to the history of our civilization. Second, we shall have to avoid resolutely any large-scale use of one major source of energy, however cheap and attractive that source

3

may seem to be—I mean nuclear energy, both fission and fusion energy as released by man—for otherwise we face the prospect of an almost certainly intolerable sequence of radioactive threats to our whole physical and spiritual environment. Finally, we will need to produce sufficient amounts of additional food to nurture the billions of human beings who will soon be living on this earth, or else we must grow accustomed to the idea that hundreds of millions of people will die of starvation each year.

It may seem incredible that three such awful problems could possess so simple and single an answer: to use the energy continually being received from the sun. But consider. Solar energy flows onto this planet every year at a rate more than ten thousand times larger than our current annual consumption of all other forms of energy. This huge flow of solar power promises to be maintained at a nearly constant rate over the years and centuries and eons of the future. Moreover, our use of this energy can avoid all thermal upset (thermal pollution) for the planet as a whole—and, if we are careful in the ways we use it, it can avoid all other forms of pollution as well. Thermal pollution of the earth as a whole can be avoided, that is, if we use the earth's currently received solar energy, because practically all of that energy is going immediately into heat in the planet's land areas and oceans and atmosphere anyway. Therefore, the total thermal balance of the earth is not changed whether or not we choose to use the energy from the sun. If we do use it, we can thereby satisfy our energy needs, and we can do so without resorting to the release of energy from the nuclei of atoms here on earth. Best of all, perhaps, solar radiation is exactly the kind of energy required to produce the vast supplies of food that we will soon be so desperate for.

So yes, the use of the earth's currently received solar energy

4

is the immediate answer to all these problems. In Chapter Eight I describe how we can use solar power—how current scientific research and engineering development work can enable us to achieve this goal, how solar energy can indeed be the *practical* and *economical* answer to our most immediate and pressing problems.

Beyond this, however, stands the fact that the use of solar energy can be the *ultimate, long-term answer* to our energy needs on earth. The analysis presented in this book shows that solar power is, must be, and will necessarily remain the source of the great majority of all the energy that we will ever find it permissible to consume on this planet for so long as the sun continues to shine at its present rate.

This fact implies a limit on our growth as a global species, of course. The rate of flow of solar energy is finite, after all, so it naturally follows that our world-wide population and human activity levels cannot grow above certain definite amounts without producing catastrophic consequences for the preservation of our civilization.

Therefore, one of the most urgent tasks awaiting us is to understand and attune ourselves, as a global species, to the facts and laws governing the way this world operates under the sun. Solar power, properly understood, can be both the immediate and also the ultimate answer to our food and energy needs, and it can do this for us without propelling us into the nightmare of nuclear energy.

The answer to these three enormous problems lies close at hand. We must demand of ourselves the wisdom to see it and develop it. The answer is to *use* the earth's currently received solar energy.

But will we see this answer? Will we be allowed to see it? Like the background chant in Greek tragedy, the phrases "energy shortage" and "energy crisis" have come to seem a permanent, numbing dirge in our lives, an ominous litany wailed to us by a chorus of governmental officials, editorial writers, and oil company presidents. We are warned to turn off that light burning in the kitchen if we are going to watch television in the family room. We are begged to correct our gasoline-devouring driving habits. We are made to feel grateful to the oil companies for their diligent efforts to drill at the farthest reaches of the earth so that we may continue to live our lives happily and comfortably as our time runs out.

These daily incantations and their intimations of impending and permanent shortages have led us to be firmly convinced that the next gallon of gas we buy may be our last, the next flick of the wall switch may bring the final sputtering spark of electricity into our homes.

Nothing could be further from the truth.

The only important kind of energy that is truly in short supply is food, the unique kind of fuel able to power the human body. While all other forms of energy customarily available to men in the past are still abundant on the planet, food is growing scarce. We can already see—we have increasingly seen over the past two hundred years—that feeding man's soaring numbers is becoming an ever more challenging task. According to some estimates, ten million people, most of them children under five, will die of hunger or of hunger-related diseases this year. In Africa, men, women, and children scratch drought-parched land for scraps to eat. In Asia, close to 750 million human beings live on diets that give each of them less than 1000 calories a day, the minimal amount a person needs if he is to have enough energy to do any sustained and useful work.

The proposition that millions around the world are malnourished or starving to death is beyond debate. But to say that mankind will soon face, not a shortage, but a surfeit of other forms of energy is to put forth a heretical theory. Admittedly, we have taken the energy available to us for granted far too long. We have been cavalier in the way we have used the earth's natural resources. We have faced, and for the next ten or fifteen years will continue to face, shortages of certain fuel supplies, long lines at some gasoline pumps, uncooled offices on a few hot summer days, some chilly homes on cold winter nights. But these problems—inconvenient, irritating, and often frightening though they are—have in fact spurred corporations, nations, and international cartels to make greater and greater efforts to supply us with an overflowing abundance of energy for the centuries to come.

Vigorous attempts are being made to find and exploit new deposits of oil. Because most of the world's abandoned oil reservoirs are still more than half full of crude petroleum, innovative techniques are being developed to draw more energy out of these long-neglected fields. The world's massive deposits of coal are being scrutinized in a new light, and plans are being laid to mine these deposits, use some of them in their natural form, and convert more of them into synthetic gasoline and synthetic natural gas. In the past, natural gas trapped in coal veins was allowed to escape into the atmosphere unused, but now efforts are being made to extract and market the huge deposits of natural gas that are part and parcel of the world's coal deposits. Tar-sand and oil-shale deposits are, for the first time, receiving earnest attention. In addition, vast commercial empires are seeking to domesticate and put to use one of the basic energy sources of the universe, the nucleus of the atom.

All these efforts are being spurred on by the demands of industrialized nations, the needs of fast-growing populations, the aspirations of people in less developed countries seeking to enjoy the levels of consumption reached in the more developed ones. I am sure that the efforts will succeed abundantly. If we are not careful, in fact, we may very well drown in this surfeit of energy.

Fossil fuels, the energy-rich materials derived over millions of years from the long-buried bodies of organisms that once flourished in the sunlight, were used for decades before we came to understand that, in addition to powering our civilization, they can also befoul our air and our water, and expose us to chemicals that cause cancer, lead poisoning, mercury contamination, and attacks of asthma and emphysema. Suddenly concerned about these side effects, we have recently launched sizable though belated efforts to reduce and neutralize the damage that fossil fuel consumption is doing to our lives.

But the recent energy crisis and our anxiety to relieve its effects, to guarantee ourselves new and ample sources of energy, are now leading us to ignore the lessons we ought to have learned about the dangers of unlimited energy consumption. Instead of learning from the recent past, we are once again rushing unthinkingly into a new and ever more massive consumption of energy. In the process, we are also hurtling toward an insidious and threatening form of pollution. Though it will be far more serious than anything we have ever confronted, we will be tempted to ignore this pollution because we will not be able to feel it, see it, taste it, or even smell it when it is wreaking its most serious damage.

This pollution is too much heat—too hot a world: global thermal pollution. If we are not careful, if we do not plan in great detail the ways in which we consume our newly developed energy sources, if we rush blindly ahead to satisfy

8

our desires, if we go on with our competitive drive to best our neighbors with energy-consuming goods and comforts we do not really need, we will release vast quantities of heat into the world's atmosphere. This heat, this thermal pollution, will warm the oceans and the air gracing the earth. In time, the heat will reach the northern and southern poles of the planet, and it will assault the massive ice sheets that sprawl over millions of square miles of Antarctica and the Arctic Region.

As these vast storehouses of ice melt, the waters will run into the oceans, swelling them and forcing them up over their shores into the hearts of most of our major cities and best farm land. If we allow this to happen, tens, hundreds, and thousands of millions of people will be displaced from their homes. Many more millions, including those unaccustomed to hunger, will starve. There will be riots and wars as those displaced by the waters try to move inland and nations fight frenziedly for the remaining food supplies.

It is controversial—exceedingly so—to warn that unless proper precautions are taken, mankind will soon be using nuclear and fossil fuels at rates sufficiently high to melt the ice caps. It is controversial because the energy consumption levels needed to bring about this catastrophe seem enormous by the standards of past and present-day experience. And many people, including many scientists, have difficulty believing that energy consumption can grow very much larger than it is today.

But I firmly believe that we can—and for the sake of our descendants, we must—foresee a world in which energy is consumed at rates far greater than today's. "If the underdeveloped parts of the world were conceivably to reach by the year 2000 the standard of living of Americans today, the world-wide level of energy consumption would be roughly

100 times the present figure," Chauncey Starr, president of the Electric Power Research Institute in Palo Alto, California, has said. "Even though this is a highly unrealistic target for thirty years hence, one must assume that world energy consumption will move in that direction as rapidly as political, economic, and technical factors will allow." Dr. Starr is unquestionably right in doubting that a hundredfold increase in energy consumption will occur within only thirty years. But he is also quite right, I believe, in saying that energy consumption will increase as fast as politics, economics, and technology will allow. These factors may not permit energy consumption to grow a hundredfold within thirty years, but there is little doubt in my mind that they *will* permit energy consumption—will even encourage it—to grow a hundredfold within eighty to one hundred twenty years, a thousandfold within one hundred twenty to one hundred eighty years, and ten-thousand-fold within one hundred sixty to two hundred forty years. A thousandfold increase in the rate of our fossil fuel and nuclear energy consumption will certainly be sufficient to bring on melting of the ice caps; a ten-thousand-fold increase would surely cause them to melt in a few decades.

What I am saying is that the current acceleration of our world-wide nuclear and fossil fuel consumption rate will, *if that acceleration is sustained at present levels for another eight to eighteen decades,* produce a *predictably inevitable* calamity. We may be able to clean up or sequester the soot from our smokestacks, the carbon monoxide from our exhaust pipes, the sulfur dioxide from our power plants, possibly even the radioactive poisons from our nuclear energy generators. But the basic laws of the physical universe dictate that practically all of the energy powering every human activity in every corner of the world must ultimately warm

10

the atmosphere.] This heat cannot be contained, it cannot be transformed into beneficial or nonthermally polluting products. The waste heat from a factory, for example, might be siphoned off into nearby homes; but after it has warmed the houses and their inhabitants, that heat must eventually find its way up into the atmosphere and then into outer space. Therefore, if we continue to accelerate our consumption of the earth's bountiful supplies of fossil fuel and nuclear energy, this ever warming atmosphere must inexorably melt the ice caps—and we must then inevitably suffer the flooding of our coastal cities and farm lands.

Some scientists are willing to acknowledge the dangers inherent in an unbridled use of the earth's energy resources. Although man's energy production is "too small at present to create a problem," says Dr. Wolf Häfele, director of the Institute for Applied Systems Analysis and Reactor Physics at the Nuclear Research Center in Karlsruhe, Germany, projected and expected energy outputs are "cause for concern." Though he does not explicitly forecast the melting of polar ice, the impact of human activities, Dr. Häfele has warned, could make itself felt in several ways. "The first effect might be the alteration of the pattern of the rainfall cycle. There is already some indication that the number of heavy rainstorms over industrial areas is changing . . . The alteration of climatic patterns over large areas . . . might be a second stage . . . A still more serious third-stage effect of man-made power densities would be on the global climate as a whole, including an increase in the average temperature. Climatic temperature changes of even 1-2 degrees would be significant."

Although we could easily slip into the thermal pollution catastrophe, neither that very possible disaster nor the very real catastrophe we are already facing—the starvation of millions of people—is unavoidable. Though they seem at first glance to be unrelated problems, both thermal pollution and global food shortages have one solution in common: the harnessing of the earth's currently received solar energy to grow vegetation on millions of acres of ocean now unused —vegetation that will provide food for mankind, and yet more vegetation that can be converted into nonthermally polluting fuels for powering man's industrial engines, for keeping him and his supplies moving, for heating and cooling his homes and businesses, for manufacturing artificial fibers and plastics of every kind. A start toward this extremely practical solution has recently been made with the "planting" of a small, experimental ocean farm near San Clemente Island, some sixty miles off the coast of California.

If we do accept the potential for catastrophe, if we pause to consider the dangers inherent in our burgeoning populations on earth, if we understand how competitive interactions among individuals and rivalries among nations spur an irrational rise in energy consumption rates, if we recognize that the world's remaining supplies of fossil fuels plus the soon-to-be-exploited sources of nuclear power can easily sustain our constantly accelerating rates of energy consumption well beyond the "flash point" at which the melting of the ice caps will start, then perhaps we will not slip into the thermal pollution catastrophe. And if we pause long enough in our contemplation of the choices before us, we can, I believe, even find a way to use our almost limitless supplies of energy, not just to avoid catastrophe, but to provide every member of the family of man with a satisfactory standard of living.

2
THE WATERS COME

On Greenland and Antarctica, at the northern and southern tips of our globe, giant ice sheets thousands of feet thick cover millions of square miles of land.

For hundreds of thousands of years, raging blizzards and temperatures as low as 125 degrees below zero have locked more than five million cubic miles of ice into the bleak expanses that dominate the top and bottom of the planet. But if man continues to accelerate his consumption of oil, coal, gas, and nuclear power as furiously as he has up to now, the heat he will produce in the process will be carried to the poles by winds and ocean currents. As the temperatures of the polar zones rise, the huge stores of ice will melt.

The melting will start slowly, almost imperceptibly. Because the ice floating on the Arctic Ocean is thin and relatively close to the centers blasting forth man's heat, this ice will no doubt be the first to melt. The direct climatic effects of that melting will be fairly small, since the sea ice constitutes less than two-tenths of one percent of the earth's total ice; but the loss of the ice will cause the whole north polar zone to increase its absorptive power for the sun's radiation, and hence it will greatly accelerate the rise of temperature in the entire region. Therefore, the Greenland ice sheet, which contains about 11 percent of the world's ice, will probably melt next. In time, if man's accelerating production of heat continues, the natural forces that control the world's climate will be increasingly influenced by this man-made warmth, and the Antarctic ice cap in turn will divest itself of its long-frozen burden of water in an avalanching rush. Indeed, my calculations strongly indicate that if man continues his exponentially accelerating consumption of thermally polluting energy at increments of 4 to 6 percent per year, the final 90 percent of the Antarctic ice will melt within a few decades.* To make matters worse, this final rush of melting may well be irreversible—beyond man's technological powers of control.

By the time the last chunk of slushy ice falls off Antarctica into the sea, the waters of the world's oceans will have risen

---

* A paper entitled "Estimates of the Magnitudes and Time Lags of the Earth's Thermal Response to Man's Total Energy Input Rates" is available from the author for the price of postage and photocopying. For detailed information write to Dr. Howard A. Wilcox, Code 0103, Naval Undersea Center, San Diego, Calif. 92132.

160 to 200 feet. They will have cascaded tens or hundreds of miles inland in many places, and will have destroyed cities, rich agricultural lands, entire countries.

The catastrophe will gather its forces bit by bit. At first, man's heat emissions into the atmosphere will draw little or no attention. Temperatures in the countryside may go up slightly. However, air-conditioning systems for homes, apartment buildings, and office-laden skyscrapers will be so commonplace and powerful in the decades to come that most people will be even more insensitive to the vagaries of weather and climate than we are today.

Some of us, no doubt, will notice that we are running our air-conditioners more in summer than we once did, or that our heat consumption in some recent winters is not as high as in previous years. A few of us may notice an increase in the number of heat waves stifling the cities. Some people may become concerned about the slightly increased number of heat-related deaths—more sunstroke, more heart attacks among the elderly and the overworked. But by and large most of us will simply go on with our daily pursuits, oblivious to the slowly changing climate.

In time, as more and more conventional power plants and atomic reactors around the world work harder and harder to keep up with civilization's soaring demands for energy, as the heat plumes escaping the vast cities all over the globe work their way up and out into the atmosphere, the omens of impending disaster will be etched more clearly.

In many coastal towns today, occasional high tides, driven by high winds, inundate water-front areas. For example, in California's Ventura County, just north of Los Angeles, tides have periodically combined with storms or unusual swell conditions to erode beaches and undermine houses built too close to the shore. When the ice caps first start melting at an appreciable rate, oceanographers will be publishing scientific papers taking official note that the seas are rising a few inches every year. Summertime beachgoers will find that winter gales and an increasing number of record tides have swept away more and more of their favorite beaches. Owners of beach houses will find that off-season storms have washed out the homes they had built years before on supposedly safe ground.

As the distant polar ice caps begin divesting themselves faster and faster of their vast burden, residents of waterfront cities like Buenos Aires and Singapore will hear a growing number of reports by television weather reporters that all-time records for high waters were set the night before. Day by day the skirmishes between the swelling oceans and the coastal cities of the world will turn into increasingly serious confrontations. Not long after new high-water marks become a weekly occurrence, the surging tides combined with severe storms will wash away buildings and docks in the water-front areas of cities like Tokyo and Yokohama. If the Amalienborg Palace in Copenhagen and the new Opera House in Sydney are still standing toward the middle of the next century, they will be repeatedly flooded during the record tides that will come more and more often as thermal pollution tightens its grip and the destruction of earth's ice accelerates.

Countries whose coastal areas are flat plains rising only a little above sea level (as distinguished from those areas where the coastline is composed of bluffs 50, 100, or 200 feet high) are likely to suffer an increasing number of disasters similar to those they experience today when great storms strike along the coasts. In Bangladesh, for example, the major rice-producing area consists of hundreds of small islands, called chars, that are nearly level with the neighboring ocean. Every day, thousands of peasants sail from the higher lands to the chars to cultivate the rice fields. During the monsoon season in November 1971, a giant storm drove waves 15 feet high across these islands and swept an estimated 300,000 people to their deaths. Soon after that disaster the Red Cross announced plans to build a series of twenty-foot-high platforms that would give refuge to the field workers if another cyclone should strike during working hours. Each platform was designed to be 100 feet wide by 150 feet long, big enough for a thousand people and their animals, and each was to hold enough food and water to sustain its refugee population until relief could be sent. These platforms may be high enough to save people today —assuming that in the face of a cyclone a thousand people and their animals could get up on one of these platforms in orderly fashion—but they would be of little use if the ocean itself were to rise a few feet.

In 1953, on the night of January 31–February 1, a raging storm allied itself with unusually high tides to send the North Sea careening across vast areas of northern Europe. Millions had to leave their homes as the waves rumbled through the Netherlands, Belgium, and England. In the United Kingdom the damage was especially serious. The sea's waters burst up the Thames River as far as the London suburbs. More than two hundred miles of land along the low-lying British North Coast were ravaged. In one rich farm area alone, more than ten thousand acres of land were ruined by the salty waters. Although scores of bodies were found— seventeen in the Norfolk area—the total number of victims was impossible to establish. More than five hundred people who had lived on a tiny island at the mouth of the Thames were missing after the storm.

17

Even though the ice at the poles will be melting with increasing speed during the early years of thermal pollution, the waters will rise slowly enough at first to allow engineers and technicians to build new and higher dikes around major population areas in coastal lands every few years. Unfortunately, these barriers are not likely to help for long, especially if storms drive mountainous waves against them. During the 1953 North Sea storm, hurricane-force winds agitated the sea so that walls of water up to 12 feet high breached Holland's famous dikes in some seventy different places. As the waters raced inland, they swept through villages and towns where people, warned by tolling church bells and screaming sirens, were vainly trying to seek shelter on rooftops, in trees, and even on the crumbling dikes. In all, nearly eighteen hundred people lost their lives, including many who died slowly of exposure while waiting for rescue.

It is impossible to foresee every detail of the disasters that will befall the world in a time of rising waters, but such disasters will certainly come if we continue the reckless acceleration of our consumption of energy, if we fail to curb our appetite for growth. We can be sure that in areas like England's North Coast, Belgium, Bangladesh, Holland, and such low-lying areas as East Texas, for instance, the swelling oceans, whipped on by storms and abnormal tides, will be merciless. Even in terms of present population, the number of deaths and the intensity of human suffering would be vast. If population densities continue to increase, as most demographers expect, the loss of life would be almost unimaginable.

All too soon the ice will be melting so quickly and the waters rising so high that, dikes or no dikes, many coastal cities will have to be abandoned. Amsterdam and Venice (if in fact Venice survives other problems that are already sinking it today) will be among the first cities to be given up completely to the seas. Bombay, for example, will probably not survive the first ten- or twenty-foot rise in sea level, because one-fourth of the city actually lies on land that is below sea level and much of the city has been built on areas reclaimed from the sea.

In cities just slightly above sea level—Tokyo, New York, Rome—the encroaching oceans will first spread just a few inches of water across the streets. This may not sound catastrophic, but even if walkways were constructed above street level, havoc in those urban areas would quickly become the norm. Major means of transportation—the subway in New York, the "tube" in London—will have to be shut down frequently, eventually almost constantly, to allow engineers time to dry out and repair the network of tunnels and stations. Nearly all the underground power and telephone cables will be breaking down repeatedly under the corrosive attack of the salt-laden water. In all likelihood, important cities will have to be abandoned even before the rising waters destroy the older buildings and the waves start to lick at the second-, third-, and fourth-story windows of the newer, sturdier skyscrapers.

Because many low-lying cities—like Washington in the USA, Cape Town in South Africa, Oslo in Norway—are important nerve centers, it is probable that attempts will be made to turn them into latter-day versions of Venice. But in time, all but the most hardy (or most stubborn) urban residents will have to leave their homes because the seas will be continuing their rise, inexorably submerging and undermining building after building.

The waters, of course, will not rise just to engulf cities like Baltimore, New Orleans, Amsterdam, Rotterdam, Antwerp, and Berlin. The waters will spread in every direction, swallowing territories far inland, especially in the fertile, near-sea-level areas where great rivers meet the oceans.

Some elevated territories along the western coast of South and North America will stay dry and safe. But the enormous Amazon Valley will, under the influence of the spreading oceans, become a gulf cutting its way through the broadest part of South America. Central and North American territorial losses will be substantial as well. The Yucatan Peninsula and a large part of the west coast of Mexico will sink beneath the waves. In the USA, all of Florida will be lost to the seas, while vast bodies of land near such rivers as the St. Lawrence, the Hackensack, the Delaware, the Potomac, the Roanoke, the Savannah, the Chatahoochee, the Alabama, the Rio Grande, and especially the gigantic Mississippi, will be lost as the oceans enter their mouths and push inland. The delta of the Colorado, as well as the valleys of the San Joaquin, the Sacramento, and the Columbia rivers, will be drowned by the time the ice caps have finished their melting.

In Europe, much of Norway will stand out of the water, but practically all of Holland, Belgium, and Luxembourg will be submerged. By the time the moving waters come to a halt along the northern part of the continent, Germany will have coasts along the greatly enlarged English Channel. Finland will have lost most of its area to the Baltic. Sweden will have lost Stockholm, and the Barents and Kara seas will have inundated hundreds of thousands of square miles of northern Russia's low-lying lands. Leningrad will be lost, and Moscow, now six hundred miles from the ocean, will find itself much nearer the shore. The waters will reach up the Seine toward Paris, up the Thames well past London, up the Tiber beyond Rome, up the Po toward Milan, up the Danube toward Bucharest.

The valleys of the Nile, the Tigris, and the Euphrates will be added to the ocean's conquests. The beds of the Brahmaputra and the Ganges will cede thousands of square miles to the seas. The Mekong Gulf will cover half of what is now South Vietnam. The fertile Yangtze Valley will be inundated by sea water.

Because the oceans will have covered substantial parts of the world's major river courses and their rich agricultural soils, there will be much less land available on which to grow food. The farm lands that survive the floods will not bear as much food as before, because most of the world's fertilizer plants—which are in and around major coastal cities —will have been quickly flooded out. Both urban and rural life will be severely disrupted because many of the largest electrical power plants supplying electricity to distant cities and farm areas will have been swamped by the rampaging oceans. Major rural irrigation projects will become inoperative because of the loss of electrical pumping power. With most of the world's important industrial areas gone, there will be little or no harvesting equipment, little or no packaging equipment, few preservatives, few insecticides. There will simply be no way to sustain an agricultural system capable of feeding billions of people; famine and starvation will become almost total.

It is quite likely, then, that by the time the waters top out, man's present civilization will have crumbled. Perhaps five to twenty-five billion people (depending on how high the world's population has soared) will become refugees in search of dry land. As they flee the coasts in pursuit of new lives farther inland, as they descend like locusts on all available space, food, and shelter, they will undoubtedly be opposed by the peoples living high and dry away from the shore lines. Riots are likely to break out within countries like the United States, where there are long coastal zones and substantial inland areas. Wars will erupt between smaller nations as people from inundated countries try to escape to neighboring lands farther from the threat of the seas.

The following table indicates some of the world's larger cities* that will be endangered if the ice caps melt:

World's Metropolitan Areas of More than 2 Million People (ranked by population as of 1964)	1964 Population (millions)	Elevation (feet)	Flooded if the Ice Caps Melt
1. New York, USA	16.325	55	Yes
2. Tokyo, Japan	15.4	45	Yes
3. London, England	11.025	80	Yes
4. Osaka, Japan	8.7	40	Yes
5. Moscow, Soviet Union	8.45	425	No
6. Paris, France	8.0	250	No
7. Buenos Aires, Argentina	7.7	45	Yes
8. Shanghai, China	7.6	20	Yes
9. Los Angeles, USA	7.475	340	Partly
10. Chicago, USA	7.09	595	No
11. Calcutta, India	6.7	20	Yes
12. Mexico City	6.1	7,349	No
13. São Paulo, Brazil	5.45	2,545	No
14. Rio de Janeiro, Brazil	5.25	30	Yes
15. Essen-Dortmund-Duisburg, W. Germany	5.2	208	Partly
16. Bombay, India	4.7	25	Yes
17. Cairo, UAR	4.6	98	Yes
18. Peking, China	4.2	165	Probably

* See also the table of countries in Notes and References, pages 157-161.

World's Metropolitan Areas of More than 2 Million People (ranked by population as of 1964)	1964 Population (millions)	Elevation (feet)	Flooded if the Ice Caps Melt
19. Detroit, USA– Windsor, Canada	4.17	585	No
20. Philadelphia, USA	4.15	100	Yes
21. Berlin, W. Germany	4.025	115	Yes
22. Leningrad, Soviet Union	4.0	33	Yes
23. San Francisco–Oakland– San Jose, USA	3.73	65	Yes
24. Boston, USA	3.48	21	Yes
25. Tientsin, China	3.4	15	Yes
26. Victoria, Hong Kong	3.275	50	Yes
27. Seoul, S. Korea	3.2	75	Yes
28. Djakarta, Indonesia	3.15	16	Yes
29. Manila, Philippines	2.9	30	Yes
30. Delhi–New Delhi, India	2.9	770	No
31. Manchester, England	2.85	275	Partly
32. Milan, Italy	2.775	397	No
33. Shenyang (Mukden), China	2.65	560	No
34. Birmingham, England	2.64	452	No
35. Wuhan, China	2.6	55	Yes
36. Madrid, Spain	2.575	2,150	No
37. Rome, Italy	2.5	66	Yes
38. Santiago, Chile	2.4	1,795	No
39. Sydney, Australia	2.34	35	Yes
40. Lima, Peru	2.3	501	No
41. Hamburg, W. Germany	2.3	20	Yes
42. Washington, D.C., USA	2.265	25	Yes
43. Budapest, Hungary	2.265	370	No
44. Cleveland, USA	2.26	580	No
45. Montreal, Canada	2.25	104	Yes

World's Metropolitan Areas of More than 2 Million People (ranked by population as of 1964)	1964 Population (millions)	Elevation (feet)	Flooded if the Ice Caps Melt
46. Johannesburg–Germiston, S. Africa	2.2	5,689	No
47. Barcelona, Spain	2.175	43	Yes
48. St. Louis, USA	2.155	455	No
49. Nagoya, Japan	2.15	50	Yes
50. Madras, India	2.15	30	Yes
51. Bangkok (Krung Thep), Thailand	2.1	10	Yes
52. Karachi, Pakistan	2.1	60	Yes
53. Melbourne, Australia	2.055	30	Yes
54. Chungking, China	2.05	787	No
55. Canton, China	2.05	33	Yes
56. Vienna, Austria	2.025	550	No
57. Tehran, Iran	2.0	3,865	No

The question, of course, is whether the catastrophe is inevitable. Must the ice caps melt? Must our cities and agricultural lands be flooded? Must there be starvation and social chaos?

The answer must be yes—if we continue to cherish the ideal of unlimited growth. Certainly—if we constantly accelerate our consumption of fossil fuels and nuclear energy. Undeniably—if we push ahead with our relentless quest for more things, bigger gadgets, more comforts, more luxuries— for consumption patterns marked, in short, by glut and the superfluous. It is under the influence of this manic greed that we may unwittingly cross over into thermal pollution, bringing disaster on so quickly and with so much intensity that it will withstand any efforts we might muster to try to avert the loss of the ice caps and the swelling of the oceans.

The catastrophe does not have to come. We can prevent this thermal pollution disaster and build a world for all to enjoy if we will but stop and question the road we are following, if we will take stock of the implications of our way of life, if we will *comprehend* that an abundant source of energy, the sun, stands ready to support our hopes with all the food and all the nonthermally polluting fuels we have the wit and the will to produce from its beneficent rays.

3

HOTHOUSE EARTH

To understand why the acceleration of man's consumption of the earth's buried resources of energy—if that acceleration is maintained for long—must inevitably lead to a melting of the ice caps, we need to understand how our planet is warmed under natural conditions, how heat acts in our fragile atmosphere, how man produces heat, and how that heat makes an impact on the balance between the world's ice, atmosphere, and oceans.

The natural heat in the surface environment of the earth— the lands, the oceans, and the atmosphere—comes almost entirely from the sun. Sunlight streams down through the atmosphere as short-wave-length radiation in the form of energy packets known as photons. As these short-wave photons strike the surface, they "explode," releasing heat in the process. As more and more photons pour in, the earth's surface heats up bit by bit.

A portion of the energy of the surface is radiated directly up into the atmosphere in the form of long-wave photons. If we could look at these photons in slow motion (it would have to be the most exquisite of slow motions, since all photons, short-wave-length and long-wave-length alike, travel at the prodigious speed of about one billion feet or 186,000 miles every second), we would see that once the long-wave photons leave the surface, they are absorbed by the lowest layer of air. That layer of air keeps, as heat, some of the energy borne by the long-wave photons, but it also reradiates some energy back down toward the surface and some other energy on up into the next layer of air above. As this process continues, in layer after layer, the heat from the surface works its way up higher and higher into the atmosphere.

Another portion of the heat produced at the surface by the short-wave sunlight which is absorbed there works its way up into the atmosphere through a process called conduction. The air molecules which are closest to the surface (mainly oxygen, nitrogen, water vapor, and carbon dioxide) are jostled when the faster moving molecules of the surface strike them and make them speed up. These molecules in turn bump other molecules in the atmosphere, forcing them to move faster. As the reaction spreads, this portion of the heat also spreads through the atmosphere. (When we measure the temperature of a substance, we are actually measuring the average speed of random movement of the jostled molecules in the substance. The faster the molecules move at random, the higher their temperature. The heat in the substance is the energy in the random motion of its molecules.)

Some heat also works its way up from the surface through convection, with the help of vertical winds that scoop up the heat near the surface and deposit it high in the atmosphere.

Finally, when water evaporates from the earth, it absorbs a relatively large amount of heat from the surface and carries it aloft. Then, when the vapor condenses high up in the atmosphere, this heat is transferred to the air at that altitude.

Almost all the heat brought up into the atmosphere by these various processes—long-wave photons, conduction, convection, evaporation and condensation—ultimately can (and must) leave the earth only in the form of long-wave-length radiation from the "top" of the atmosphere.

Glass covering a hothouse keeps the inside of the structure warm, because the glass allows the sun's incoming short-wave radiation to pierce it easily but does not allow long-wave radiation produced within the hothouse to emerge nearly so easily. Glass is almost transparent to short-wave photons, but it is almost completely opaque, impervious, to the outgoing long-wave-length radiation. Like the glass around a hothouse, the atmosphere is nearly transparent to short-wave radiation but nearly opaque to long-wave radiation. Only 5 to 10 percent of the long-wave radiation manages to dart directly from earth to outer space; the other 90 to 95 percent takes a laborious path up from earth toward space. Because the atmosphere produces a "hothouse effect" by acting like the glass covering of a hothouse, because it traps heat and holds it captive in a ricocheting system before it allows it ultimately to radiate away, the temperatures at sea level are some 50 to 70 degrees higher than those at the top of the atmosphere.

The shifting temperatures of the earth-atmosphere system are determined by the flows of heat within its parts. For example, as more and more of the photons from the sun flow into the northern hemisphere after the vernal equinox, the surface and lower atmosphere temperatures there are raised higher and higher. Therefore, heat is increasingly driven from this higher temperature region toward the lower temperature regions of the southern hemisphere, the upper atmosphere, and outer space. Looking at the globe as a whole, and averaging over months or years of time, we see that heat is forced to flow up through the atmosphere from the higher temperatures below to the lower temperatures above, and energy is forced to radiate away from the "top" of the atmosphere because its temperature is higher than that of outer space. The flow of energy up through the atmosphere must be nearly equal to the flow of radiation out into space, and each of these flows must be nearly equal to the rate at which energy is being received from the sun, because otherwise the yearly average temperatures of the various parts of the system would be observed to be changing rapidly, by large amounts, instead of very slowly and slightly, as actually occurs. In other words, the annual average flows of incoming and outgoing energy have reached equality, an equilibrium level, in much the same way that the incoming and outgoing flows of water in a filling but old and leaky bathtub will reach a steady level. As water begins to pour steadily into the old tub, it begins to escape in dribbles through openings worn into the material around the stopper. At first, as the water pours in through the spigot, the water level is low and the water escapes very slowly. But as the water level in the tub rises, the pressure of the water around the stopper increases, so that the water is forced out more rapidly through the leaks. When the water level has risen to a point where the rate at which the water is being forced out is equal to the rate at which water is coming in, the water level stops rising. It has reached its steady state, or equilibrium level. If the tap is then opened wider, water suddenly starts coming in at a faster rate than it has been leaking out, and the water level starts to rise again. Increasing pressure is then felt around the stopper, the water outflow rate begins to rise, and soon a new and higher equilibrium level is reached. If the tap is next turned down suddenly, the water level begins to fall until a new, lower equilibrium level is attained.

In the atmosphere, the incoming solar heat is analogous to the water pouring into the tub; the radiation of heat away from the top of the atmosphere into outer space is the "leak" out of the tub; the level of the water in the tub corresponds to the level of the temperature of the lower reaches of the atmosphere; and the pressure around the stopper is analogous to the temperature at the "top" of the atmosphere. As long as the upper portions of the atmosphere stay at their given equilibrium temperature, so will the lower reaches of the atmosphere. The fact that the incoming solar radiation and the outgoing long-wave radiation have come to a standoff puts a lid on the state of the temperature down below.

For most of the past five hundred million years this process has maintained the earth's temperature at a level that has kept the globe completely free of ice caps at both poles. But lately—in the past million years or so—things have changed.

Common experience tells us that when we alter the steady flow of such things as water or electricity or heat, we may reach a point at which violent oscillations can strike the very thing we are trying to control. If we slowly open and close a water faucet, we may suddenly hear a violent hammering noise as the flowing water begins to oscillate, moving back and forth in the pipes.

About sixty million years ago the temperature of the earth was apparently dropping slowly. Then, about one or two million years ago (at the beginning of the age called the Quaternary), the temperature curve of the earth began to show the variations typical of a system that has entered an oscillatory state. About every 100,000 to 200,000 years since then, the temperature of the earth has cycled through one complete oscillation, first dropping by some 20 to 40 degrees Fahrenheit and then rising again as much as it had dropped. There have been several such major oscillations—plus a larger number of smaller oscillations—over the past million years, and each has been accompanied by a surge of the ice caps down toward the equatorial zones of the earth, followed by a retreat of the ice back toward the poles.

Scientific theories about the origins and causes of the earth's ice ages during the Quaternary are plentiful and varied enough to give one a healthy appreciation for the difficulties that climatologists face. Most experts agree that variations in the rate at which solar energy is absorbed by the earth is the root cause of the glacial surges, but one school thinks these variations may originate in processes occurring in the sun, and another that they may result more from small changes in the orbit of the earth about the sun. Still another group of scientists thinks the variations may be produced by changing amounts of carbon dioxide or water vapor in the earth's atmosphere, and a fourth group suggests they may be caused by dust kicked up during periods of great volcanic activity or derived from meteoroids falling into the atmosphere from outer space.

One reason why climatologists have been able to generate so many different and yet plausible theories of the origin of the ice ages is that (as nearly all experts do agree!) the present climatic condition of the earth makes it extremely sensitive to slight changes in the rate at which energy is

31

turned into heat anywhere in the environment. This sensitivity forces climatologists to consider many different factors that may be singly or jointly responsible for causing the observed changes in global temperatures and the size of the ice caps.

The earth's sensitivity to such changes results from the operation of several "positive feedback" mechanisms in the planetary "climate machine." One such effect may be called the ice-cap-reflectivity feedback mechanism; it refers to the fact that if the temperature of the polar zone fluctuates for any reason, and if this fluctuation is, say, in an upward (warmer) direction, then the ice caps will melt a bit. Therefore they will cover a bit less area, their reflecting power will become a bit less, more of the incoming photons from the sun will be absorbed, more solar energy will be released into the earth's local surface environment, and consequently the rise in temperature of the polar zone will be even higher than the original fluctuation would have produced all by itself.

A second positive feedback connection in the climate system may be called the carbon dioxide–ocean temperature mechanism. If the temperature of the ocean rises a bit, as a result of some fluctuation or other, then the laws of nature state that the warmer surface waters must throw off some of their absorbed carbon dioxide into the air. This increases the opacity of the air to the outgoing long-wave radiation from the surface, hence the hothouse effect is increased in strength, and therefore the rise in temperature of the earth will be a bit more than the original fluctuation would have produced all by itself.

This talk of positive feedback mechanisms operating in the climate machine would be seriously incomplete if we failed to include the water vapor–ocean temperature feedback mechanism. For example, if a temperature fluctuation makes the ocean slightly warmer, more water will evaporate into the air. The atmosphere will consequently become more opaque to the outgoing long-wave radiation from the surface, the hothouse effect will again be enhanced, and therefore the rise in the earth's temperature will be a bit more than the original fluctuation would have produced all by itself.

Of course, these positive feedback effects work just as effectively in the downward or lower temperature direction as in the upward or higher temperature direction. These effects are what can make the natural climate oscillate up and down instead of going always in one direction. But humanity's steadily accelerating consumption of power is pushing the system in only one direction—that of melting the ice caps. If man's thermal pollution should drive the climate beyond the point of no return, the melting of the ice caps would be the single and uniquely possible result.

There are also many negative feedback mechanisms operating in the climate machine. For example, if the heat flowing upward through the atmosphere were to increase a bit, as the result of some fluctuation or other, then the top of the atmosphere would tend to get warmer; it would then tend to radiate more energy out into space, and this in turn would tend to cause the top of the atmosphere to cool off, thereby diminishing the effect that the original fluctuation would have produced all by itself. As another possible example, if a rise in surface temperature were to increase the amount of low-lying cloud cover around the earth, this might increase the reflectivity of the planet to the short-wave photons from the sun and hence diminish the magnitude of the temperature rise that would otherwise have been experienced.

So powerful are the positive feedback mechanisms operating in the climate machine, however, that some climatologists suspect they may, all working together, actually out-weigh the total effect of the negative feedback mechanisms also operating in the climate system, thus creating an un-stable regime. If this were indeed the case, then the earth's climate would be an "oscillator." Even if the energy emitted by the sun were to stay strictly constant (which, of course, it does not), the earth's climate would then either show a cyclic oscillation between extremes of temperature and glaci-ation, or avalanche in one direction toward a completely ice-free world, or slide with accelerating speed toward a heavily iced (possibly a totally ice-covered) world.

Even if the climate machine is not an oscillator, but in-stead is stable to some degree, most climatologists agree that the positive feedback mechanisms operating in it are so strong that it can at the very least be seen as a powerful "amplifier"—a machine that increases the magnitudes of the effects produced by the small variations in whatever energy is released by the sun and by man into the surface environ-ment of the earth.

The theories and measurements of the climatologists are not yet sufficiently complete and refined for them to know whether our climate machine is an oscillator or an amplifier. But whatever the reasons behind the major temperature fluctuations that have occurred over the past million years or so, it is clear that at all times during each fluctuation the atmosphere was very nearly in a heat flow equilibrium. The oscillations themselves, after all, were spread over tens of thousands of years, ample time for incoming solar energy and outgoing earth radiation to come to a balance.

It is also clear that today the earth's temperature is such as to keep the ice caps contained at the north and south poles. Although the ice caps thaw a bit each summer and refreeze again in winter, the global climate now allows the ice caps neither to melt completely nor to spread significantly beyond their current borders. It is quite possible—in fact highly probable—that left to itself, the situation now reigning would remain about the same for hundreds or thousands of years longer, and that the ice caps and their vast load of solidified water would stay just about where they are.

Unfortunately, that equilibrium may not stay at its present level much longer—not because some cosmic change is likely to affect it in the near future (let's say the next several thousand years), but because men may soon get in its way.

Although we talk about the sun's energy in highly abstract fashion, scientists have measured that energy in very real terms using various standard units of energy. One such standard, the British thermal unit (Btu, for short), is the amount of heat any source of energy—wood, food, coal, oil, the atomic nucleus, the sun—must generate to raise the temperature of one pound of water one degree Fahrenheit. Energy experts have calclulated that the sun pours the equivalent of about 5,300,000,000,000,000,000,000* Btus of energy onto the top of the earth's atmosphere every year.

Everything we do takes energy, whether we produce weapons for war or toys for children, whether we heat private homes or air-condition towering skyscrapers, whether we transport food from farm to city, workers from home to factory, or executives from city to city. And the overwhelming bulk of that energy today is produced by burning fossil fuels to generate heat: heat to drive the car, heat to propel

---

* The number in the text—5,300,000,000,000,000,000,000—can be written as 5.3×10^{21}. This means that the first number, 5.3, is multiplied by the second number, 10^{21}, which in turn can be visualized as a one followed by 21 zeros. $5.3 \times 10^{21} = 5.3 \times 1,000,000,000,000,000,000,000 = 5,300,-000,000,000,000,000,000$.

the airplane, heat to smelt the copper, heat to cook the food.

All the heat manufactured during the conversion of fossil fuels (and nuclear energy) must find its way into the atmosphere. In very small quantities it is not particularly bothersome. At present, for example, man is pouring heat into the atmosphere at a rate just a bit less than 1/10,000th of the rate at which heat is being contributed by the sun. Thus the overall temperature equilibrium—which is, of course, dominated by the amount of energy coming from the sun—is not greatly disturbed, probably not even to a measurable degree. A man-made equivalent of 0.01 of 1 percent of the sun's energy is not likely to have any measurable impact on the world as a whole.

But man's unremitting pursuit of ever higher levels of consumption—not just in the countries where there are already more cars than families, but in the developing nations as well—is leading to the exponentially spiraling use of energy. The burning of fossil fuels and the expanding use of nuclear energy promise, within the next century, to release increasing amounts of man-made heat into the atmosphere. If we keep accelerating our energy consumption in the future as we have in the past century, then by the period 2015 to 2035 we shall be adding heat equivalent to 1/1000th of the amount of heat contributed by the sun; by 2055 to 2095, man will have increased his heat output another tenfold and will be adding 1/100th as much heat to the atmosphere as the sun does. If present energy consumption trends continue, by 2095 to 2155—120 to 180 years from today—man's contribution of heat to the atmosphere will have increased another tenfold; he will then be pouring heat equivalent to ten percent of the sun's contribution into the world's atmosphere. By 2135 to 2215, man would be adding as much heat to the atmosphere as the sun itself. Indeed, if it were then possible for man to continue his present energy consumption growth pattern for still more years, he would, by 2175 to 2275, be pouring *ten times more* heat than the sun does into the earth's surface environment.

My calculations (see note page 14) show that when man's contribution of heat to the atmosphere climbs above 1 percent and begins to mount toward 10 percent of the sun's input, he will have crossed a dangerous threshold. He will be entering the realm of thermal pollution.*

If and when man begins to add heat to the atmosphere at a significant fraction of the solar rate, a host of mechanisms will come into play to bring about the rapid destruction of the ice at the poles. First will come the destruction of most or all of the ice floating on the Arctic Ocean, as convective

---

* The definition of the word "pollution" is still controversial among scientists. A good definition, I think, and the one adopted in this book, is that proposed by a panel of the President's Science Advisory Committee in 1965: "Environmental pollution is the unfavorable alteration of our surroundings, wholly or largely as a by-product of man's actions, through direct or indirect effects of changes in energy patterns, radiation levels, chemical and physical constitution, and abundances of organisms. These changes may affect man directly, or through his supplies of water and of agricultural and other biological products, his physical objects or possessions, or his opportunities for recreation and appreciation of nature." Thus pollution is a change for the worse in a complex environment in which man has been able to develop as a highly diversified species.

It is worth noting that man's addition of heat to the atmosphere, if it were to melt the ice caps over several thousand years, would not necessarily be pollution in the sense of this definition. It would not be pollution for two reasons: First, if man were given a sufficient amount of time, he might adapt quite comfortably to changing shore lines—indeed, this has happened in the past and is still happening gradually today. After all, most of our cities are old and decrepit, and most buildings are depreciated within fifty years of their construction anyway. An event that would force man to rebuild his cities might not be detrimental *per se*. Second, it might very well be that a world without ice caps, a world in which the specter of periodic ice ages is absent and which has larger tropical and temperate zones, would provide man with a better environment than he enjoys today.

In other words, thermal pollution is a catastrophe only if it engulfs man with problems faster than he is able to adapt to them.

winds and ocean currents carry man's increasing heat output into the polar environment. The Arctic Ocean ice, which is only five to ten feet thick, will probably melt within five to ten years. This melting itself would not produce a rise in sea level, because the ice is already floating and has already produced a rise in the ocean level to support its weight. (Put an ice cube into a glass of water. Immediately, the water level rises. As the cube melts, the ice disappears, but the water level does not thereafter rise beyond, nor fall below, the point it reached when the cube was first put in.) Today, however, the ice of this northernmost area acts like a mirror, reflecting most of the sun's incoming short-wave radiation back into space before that radiation can be converted into heat. The melting of the Arctic Ocean ice will strongly decrease the ability of the north polar region to reflect this incoming solar radiation back into space. Therefore more solar radiation will be absorbed and more heat will be formed in the Arctic Region. Hence the ice sitting on the huge island of Greenland will melt faster. The melting of the Greenland ice will itself raise the world's ocean level by some 20 feet or so. Once the Greenland ice cover has melted, the real assault on the Antarctic ice sheet will begin. As the north polar region warms up, the world's climate machine will drive more heat toward the Antarctic. Since the ice sheet covering this continent contains some 85 percent of the world's total ice, its melting will raise the level of the oceans by an additional 140 to 180 feet.

An increase in the earth's temperature—even a slight increase—could strongly gnaw away at the ice covering the poles. The ice caps, after all, are exceedingly fragile formations; they are not like rocks or metal. M. I. Budyko, the world-famous Soviet expert on the earth's climate, has estimated that if the energy input to the earth's surface were to be raised, on a sustained basis, by only 1 percent or so of the solar input rate, a complete melting of the polar ice would occur. The noted American climatologist, Professor William Sellers of the University of Arizona, has come to similar conclusions. We know by scientific observation that the area covered by north polar ice did actually diminish temporarily in the 1920s and 1930s, when the global average temperature increased briefly (no one knows why) only a few tenths of one degree Fahrenheit.

We should remember too that the ice caps are likely to melt quickly once thermal pollution starts, because man will not—at least not unless he is forced by painful consequences —level off his energy consumption when he starts to produce heat in an amount ranging from 1 to 10 percent of the incoming solar radiation. Rather, he is likely to go on consuming energy at a spiraling rate of 4 to 6 percent a year, doubling his heat production every twelve to seventeen years. Forty to sixty years after he has reached the 10-percent level, then, he will possibly be adding as much heat to the atmosphere as the sun does.

If we compare the melting process to a ball on a hill, we might say that when man's contribution of heat to the atmosphere reaches about 1 percent of the incoming solar energy (at which time he will have actually increased temperatures around the globe by only one or two degrees Fahrenheit), man will have started the ice to melting in earnest. The ball will be well on its way toward rolling downhill. If the rate of use of energy were to be leveled off at that value, the ball would probably keep on going, but at a comparatively slow speed. The melting might proceed at a speed slow enough to stretch the process out over hundreds of years, enough time for man to adapt reasonably quickly. But if he goes on with his 4 to 6 percent yearly increase in energy consumption, man will, having launched the ball downhill, be running along behind kicking it faster and faster. He will be giving the caps no quarter. Thus he will guarantee their rapid destruction, and probably the obliteration of his present civilization as well.

4
THE WORLD FOOD CRISIS

Possibly we may be able to revel in our glut of energy for several more decades before we are forced finally to pay attention to thermal pollution and to the problems we engender as we go on increasing our use of fossil fuels and nuclear energy. But if we continue to cling to our irrational ways of using the earth's natural resources, we will soon be forced to cope directly with another result of our actions: the starvation, year after year, of hundreds of millions of men, women, and children around the earth.

Hunger has haunted man for hundreds of years. In the time that has elapsed since Thomas Robert Malthus first predicted that the world's population would always increase faster than mankind's ability to grow food, economists, agronomists, and politicians have watched as food continuously eluded the reach of millions of people. According to Dr. Philip Handler, president of the National Academy of Sciences, there were two million hunger-related deaths in the seventeenth century, ten million hunger-related deaths in the eighteenth, twenty-five million in the nineteenth. Now, in the midst of man's richest times, mass starvation is about to increase even more dramatically—and it seems likely to go on doing so well into the twenty-first and twenty-second centuries. Food is so scarce, says the Food and Agriculture Organization of the United Nations, that 33 to 50 percent of the people of the world—1.3 to 2 billion men, women, and children—are undernourished. Many food experts estimate that hundreds of millions of people will starve to death before the end of this century, and that billions more will die in the next, unless we take radical and imaginative steps to provide food for them.

It is the children who suffer most. According to the most expert and reliable estimates, tens of millions of children are going to bed hungry every night. More than five million children under five years of age are starving to death each year at the present time, and the problem appears certain to get worse rather than better in the near future. Some scientists are even more pessimistic. Dr. Handler points out that in South Asia populations are soaring even though "there is no new marginal agricultural land to be opened in the region and there seems little possibility of expanding indigenous food production at an adequate rate." As a result, Handler said recently at the Annual Convocation of Markle Scholars, "without unprecedented, massive external food assistance, one must anticipate tens of millions of child deaths in that region in the next two decades and hundreds of millions of such deaths thereafter."

To some extent, the current food shortages are due, almost paradoxically, to rising standards of living among many of the world's people. The rich can and do outbid the poor for the world's food. As incomes grow, people demand not only greater varieties of food but also more high-protein foods. And to raise and fatten enough animals to meet the rising demands for protein-rich foods, prodigious amounts of grain must be used. Dr. Norman E. Borlaug, Nobel laureate in the area of food and nutrition, estimates that for every portion of beef raised to bring prime ribs or steak to someone's table, eight portions of grain must be fed to the animal, eight portions that otherwise could have provided food for as many human beings. As a result, the average American now consumes, directly and indirectly, some 1800 pounds of grain each year, whereas the average citizen of the lesser-developed nations consumes only about 400 pounds—less than a quarter as much.

Dr. Lester R. Brown, Senior Fellow with the Overseas Development Council, writes in his recent book, *By Bread Alone,* that the increasing pressure on the world's granaries does not come from affluent North America alone. "In the northern tier of industrial countries, stretching eastward from Britain and Ireland and including Scandinavia, Western Europe, Eastern Europe, the Soviet Union and Japan, dietary patterns are now more or less comparable to those of the United States a generation ago. Rising incomes in these countries is being translated into additional demand for livestock products but few can respond to this growth in demand entirely out of indigenous resources. Most must import either some livestock products or the feed grains and soybeans to produce them. By 1974, at least one third of the grains produced in the world, over 400 million tons, were being fed to livestock. The global use of grains for animal feed grew by more than six percent annually throughout the 'sixties." According to Dr. Brown, the world grain supply is increasing by 30 million tons a year. But of this amount, only 22 million tons meet the food needs of the world's growing population. Eight million tons—almost one-third of the additional amount of grain the world's farmers produce every year—are heaped into the troughs in feed lots where meat animals are being raised to satisfy the rising demand for steaks, chops, and roasts.

Growing affluence in many nations accounts for some of the pressures being exerted on the earth's food supplies. But, even more important, if there is less and less food to go around every year, it is because every year there are more and more people—almost 100 million more every year—to be fed.

Teeming and expanding populations are a recent thing for man. For much of humanity's history on earth, his population was small. Like all animals, man has an instinctual desire to reproduce and generate offspring. And like all animals, he was once subject to a host of dangers that kept his numbers in check. Starvation, beasts of prey, storms, epidemics, cold, high infant and maternal mortality rates, among other things, kept human ranks thin. Thirty thousand years ago, some historians calculate, there were probably no more than one million people scattered about the globe.

As man began to establish some control over his precarious life, as he began to understand a little something about illness and to devise rudimentary cures for some of the things that ailed him, his numbers began to creep upward. By the middle of the seventeenth century, some 500 million people inhabited the earth. By about 1820 world population had reached one billion.

During the past two hundred years—and especially during the past seventy-five years—modern technology and a rapidly improving medical science have given man an even greater ability to mitigate the natural disasters that once kept his ranks thin year in and year out. Improved medical practices like vaccinations, new drugs like antibiotics and antimalarial compounds, public health centers, and the development of insect and bacterial control techniques came to mean dramatically extended life spans for all people. In addition, these advances made it possible for more women to survive the birth process and for more infants to survive childhood. Once a family might have had to produce ten or fifteen children in the hope that a few would survive. As medical science progressed, families began to have five and six children, nearly all of whom lived to be adults and have children of their own.

Numerically, these developments have meant that mankind's death rate has been lowered to about 1.4 deaths per year per 100 persons. At the same time, mankind has kept its birth rate up to roughly 3.4 babies per year per 100 people. On a net basis (subtracting 1.4 from 3.4) this means that today man is increasing his population every year by some 2.0 percent—two more people each year per 100 persons in the population.

All these percentage changes seem modest and even insignificant. But they represent huge and rapid changes in the growth of the world's population. Although it took mankind thousands of years to reach a census of one billion people, it took only an additional one hundred ten years—from 1820 to 1930—for mankind to spawn its second billion. It took just another thirty years—not one hundred ten—for man to add his third billion. Only fourteen years later he added the fourth billion.

Today's total human population is almost exactly four billion. If man's numbers had grown by the small rate of 2 percent per year for all his thirty thousand years as a domesticated creature, his numbers today would be so great that the bodies of people, stacked and packed like sardines, would cram every inch of space in the entire visible universe. At present, the 2 percent per year population increase is producing people so rapidly that if some great catastrophe were to eliminate all the people living in the United States, the rest of the world would generate enough new persons to both maintain its own population and repopulate the USA within less than three years. At the present 2 percent annual growth rate, man's numbers by the year 2100 could very well reach a total of some 50 billion. Thereafter, at the same growth rate, his ranks would swell by more than *one billion people every year*. He would be adding the equivalent of roughly one of today's Chinas, or about five USAs or two Europes, to the world every year.

Demographers disagree sharply when they discuss the ultimate level that world population will reach. Some point out that population growth, particularly in the industrialized nations, has slowed down. In some instances, the optimists say, birth rates have fallen dramatically enough to help some nations attain the magical level of zero population growth. Thus some experts insist that despite the recent surges in population, and despite sustained population growth in the lesser-developed nations, earth's human population will eventually level off when it reaches about six billion people. Others see a higher, though not necessarily distressing, population plateau. "The way I figure it," says one, "we will level off at a population of about 20 billion. The world can handle that, and we can all live very well."

The world may, of course, top off its population at 20 billion. But all history is against it, and similar predictions have been proved wrong time and time again. In 1954, for example, the United Nations was happily estimating that by 1980 the world's population would probably be no bigger than 3.3 to 4.0 billion people. Most likely, UN demographers were saying, the world's population would be about 3.6 billion. Much to their embarrassment, the world's population reached 3.3 billion in only eleven years, fifteen years ahead of schedule. As a result, the UN spokesmen decided that by 1980 there would be about 4.3 billion people on earth. In the span of only ten years the oracles had to revise their estimates upward by almost 20 percent. I think that even this projection will prove to be too conservative.

In 1957, a group of California Institute of Technology scientists, meeting to discuss the prospects for the world in the next hundred years, estimated that by the year 2000 there might be no more than seven billion people on earth. Ten years later, in a follow-up survey, the Caltech group retracted its optimistic forecast. "Projections made on the basis of plausible extrapolations of the rate of population increase indicate that the world population might well grow to about 7.5 billion by the turn of the century," Dr. Harrison Brown, one of the Caltech scientists, concluded in 1967. "It now appears that we will be fortunate indeed if we succeed in stabilizing the world's population at 15 billion persons. It is even possible that in another 90 years, the population of human beings will be approaching 25 billion people."

There is indeed very little reason to be optimistic. Even if people in the industrialized countries were to try faithfully and effectively to practice birth control, and even if each set of parents were to start immediately to have only two children —just enough to reproduce themselves—then the population of these countries would still be likely to double within twenty-seven to thirty-five years. (This may seem to be a contradiction, but a policy of zero population growth can for the moment affect only those women who are now ready and able to have children. Today there are more female children than there are fertile women. And when these additional children mature, they will also have children—they will reproduce themselves. Even if *they* stop at two, their large numbers will have been sufficient to double the present population.)

We also have to face up to the fact that population control is not a universally accepted ideal. In the lesser-developed nations, death rates continue to plummet as improved health practices make gains among the people. Population growth rates exceeding 3 percent have been reached in Latin America and some parts of Africa and Asia because formerly endemic diseases like malaria have become all but non-existent. Rationally, people in these areas should be curtailing their birth rates because they no longer need to mass produce children in the hope that one or two may survive and help support the family. Unfortunately, the experience that disease will not wipe out most of a family is still too new. The sexual drive combines with desire for many children to overcome more reasonable inclinations to stop reproducing prolifically.

American politicians and sociologists, horrified by rising population levels around the world, fly all over the globe to preach population control as if they were bearing a new message eagerly awaited by all. In truth, however, the pleas fall not just on deaf ears but often on hostile ones. In early November of 1974, Pope Paul VI delivered a blistering speech in which he denounced the industrialized nations for the efforts they were making to convince the lesser-developed nations to control birth rates among their peoples. The Pontiff told two thousand delegates, observers, and their families attending the World Food Conference in Rome that the rich nations were in fact conspiring to try to impose birth control on the poor nations of the world to keep them in their place. "It is inadmissible that those who have control of the wealth and resources of mankind should try to solve the problem of hunger by forbidding the poor to be born or by leaving to die of hunger children whose parents do not fit into the framework of theoretical plans based on pure hypotheses about mankind's future," the Pope said.

"In times gone by, nations used to make war to seize their neighbors' riches. But is it not a new form of warfare to impose a restrictive demographic policy on nations to ensure that they will not claim their just share of the earth's goods?"

We must recognize that the seventy-eight-year-old Pontiff's thoughts and feelings are shared by many people around the world.

Many if not most of the lesser-developed nations—nations that already contain two-thirds of the world's population—are apprehensive and distrustful of efforts by the rich to impose population growth limits on them. Many nations that lean toward the leadership of the Communist bloc are persuaded that population control is just an imperialistic propaganda message, a notion dreamed up by the capitalists to divert the attention of the poor from their real problems: capitalism and capitalistic exploitation. People could have all the children they want, this thinking proposes, if wealth were to be distributed properly. Moreover, most of those countries that do not want to align themselves with the Communist world and its theories probably resent with equal vehemence the proddings they receive from the rich to bring their population growth under control.

If we cannot stem hunger and starvation by putting a tight lid on our population growth, then we will surely have to find ways of enormously increasing the supplies of food necessary to sustain—and, we must hope, sustain well—all the people who will inhabit the globe.

Many agricultural experts insist on believing that this can be accomplished if we concentrate on increasing the supply of fertilizer (and decreasing its cost); if we improve irrigation techniques, to bring water to land where little or no water is now available; if we increase the use of the new, highly productive hybrid strains of wheat and rice; if we expand the amount of land under cultivation. There is more than enough land on which to grow the required food, they say optimistically. Sprawling virginal regions like the Congo and Amazon basins, they point out, are just waiting for the tractor and the combine. The President's Science Advisory Committee estimated in 1967 that the world's total potentially arable land embraces about eight billion acres— slightly more than twice as much land as we cultivate today, and almost three times as much land as is actually used for food production in any given year.

But it is extremely doubtful that these essentially traditional approaches to augmenting food production will be able for long to meet our increasingly urgent needs. Conventional fertilizers—many manufactured from oil—have soared in price and have grown far too expensive for millions of small farmers in the poorer countries. It is not very likely that their price will come down again.

It is true that additional irrigation could make some land more productive. But there is a limit to what irrigation can accomplish. India and Pakistan, for example, have managed to cultivate additional land by drilling thousands of new wells in the past few years. But at some point the drilling will have to taper off and stop. The water table that feeds these wells is dependent on underground streams and rain for its supplies. The amount of water that comes into the underground water table is limited, and on a long-term basis little or no more water can be drawn from this underground source than comes into it. Moreover, as Dr. Brown points out, in many areas land is currently made productive by irrigation systems that draw on underground water tables which are not recharged at all. When the water in these aquifers runs out, either new sources of irrigation will have to be found or the land will have to be abandoned. In West Texas, Dr. Brown adds, a million acres of land now under cultivation may eventually have to be abandoned because fossil water reserves (as these nonrenewable water tables are called) are being exhausted.

If we do make a concerted effort to increase the supply of fertilizer, and if we do succeed in irrigating previously dry lands, we may find in time that we are hoist by our own petard. The production of sufficient fertilizer to enrich enough land to feed billions of additional mouths, and efforts to irrigate arid lands by bringing in desalinated water (one of the leading contenders for new irrigation schemes), will add tremendously to the heat burden—and therefore to thermal pollution—of the atmosphere. "The production of nearly all of the world's nitrogen fertilizers (which accounts for roughly half of all fertilizer used) . . . is energy intensive, requiring large amounts of electrical power," Dr. Brown writes. "Glib talk about desalting ocean water for use in food production is particularly unrealistic," adds Georg Borgstrom, professor of food science and nutrition at Michigan State University, because, among other things, the movement of water up and over the continental expanses would require "enormous amounts" of energy. Thus, even if we succeed in increasing massively the production of fertilizers and the desalination of ocean water, we may eventually be faced by a cruel choice: Either we stop producing fertilizers and irrigation water from the ocean, thereby dooming millions to starve, or else we go on raising food with the help of these methods and add substantially to our assault on the ice caps.

Nor has the much-vaunted Green Revolution, the introduction of new hybrid strains of rice and wheat, given us enough reason to hope that truly significant inroads will be made against hunger and starvation with traditional methods. When these new hybrid plants were first introduced a few years ago, it was thought that they would increase per-acre yields so much that food shortages would be alleviated for years to come. Countries that had just barely been able to feed their own populations made plans to become major food exporters. Countries in which chronic food shortages and widespread starvation had been the rule expected to achieve food self-sufficiency with the help of the new strains of rice and wheat. But now the Green Revolution has lost much of its bloom. Populations have zoomed as agronomists realized that the new strains of crops, though more productive, are also more demanding. Unlike conventional rice and wheat, they need more fertilization and more irrigation to achieve their advertised yields. Because of their restricted genetic diversity, they are more vulnerable to pests and diseases. They have to be nurtured with the aid of more pesticides and with greater skill than the older, less productive strains. The Green Revolution has turned out to be beyond the financial and educational reach of the subsistence farmers who were to have been its prime beneficiaries.

"On a technical level [the Green Revolution] has achieved much," writes Geoffrey Barraclough. "In India alone it made possible an expansion of wheat production from 11 million to 27 million tons between 1965 and 1972 . . . But instead of producing a general improvement of living standards, it is generally agreed, the benefits have flowed to a privileged minority. It is the rich farmers who can afford chemical fertilizers, agricultural machinery, and the rest [who benefit], not the 70 percent of the poor peasants with less than an acre of land each. Moreover, it is much easier for rich land owners than it is for small farmers to get bank credit with which to carry out irrigation programs and build up large mechanized agricultural estates."

It is also self-deceptive to look around at the vast areas of uncultivated land on the globe and believe that they can be turned into rich farm lands. Dr. Brown points out that 70 percent of the land that is not considered arable today is in large part desert—the Sahara in Africa, the Thar Desert on the Indo-Pakistani subcontinent, the Gobi in China and Mongolia, the arid interior of Australia, sizable portions of Southwestern United States, even some parts of Europe, Peru, Brazil, and Central America.

In 1967 the President's Science Advisory Committee stated that most of the world's potentially cultivable land is in the tropics, but that half of that land is "inherently low in capacity to supply plant nutrients." Dr. Brown agrees: "It would be a serious error to view Africa and Brazil as vast, unexploited repositories of good farmland. Much of the potentially cultivable land is in the tropics, and experience indicates that the farming of tropical soils is often not economically feasible. The soils (which are not usually very fertile to begin with) and their protective forest cover form a fragile ecological system. Organic materials in the soil decay very

53

rapidly in the tropical climate, and the soils often lose whatever fertility they had once the forest above—the abundant
source of new vegetative matter—is removed. Thus farming
new soils may require heavy, continuous applications of
chemical fertilizer. In addition, when they are fully exposed
to sun and oxygen, some tropical soils undergo chemical
changes and compaction, becoming too solid to farm."

When agricultural optimists see vistas of new lands for
the production of more and more food, they sometimes for-
get that although some new land may be made to yield food,
a good deal of formerly productive land is continuously
being lost to the world's farmers. In almost every country,
Dr. Brown points out, the total land available for farming
decreases very year as urban areas, airports, shopping centers, and housing developments continue their inexorable
spread. "In the United States," Dr. Brown writes, "farmland
has been used indiscriminately for other purposes with little
thought given to the possible long term consequences. . . .
In Japan, the cultivated area was greatest around 1920, and
it has declined substantially since then. Some countries in
Western Europe—notably Sweden, Norway, Ireland and
Switzerland—have been losing agricultural land to urban
areas for the last several decades."

Where urban sprawl has not claimed previously useful
farm land, erosion and ecological disaster often have. "Erosion of the topsoil has reached seriously damaging dimen-
sions on around 15 percent of the world's agricultural land,"
Dr. Borgstrom writes. "Another 15 percent suffer from erosion coupled with critical mineral depletion. In many countries these losses are even higher. In the Philippines 75
percent of the farmland is regularly damaged by erosion
from torrential rains. In Somalia nine-tenths of all farmlands
are reported to be eroded or threatened. More than half

54

of India's topsoil is affected by such serious erosion that within 25 years the topsoil will have vanished on one-fourth of the tilled land." And, according to Dr. Brown, "literally millions of acres of cropland in Asia, the Middle East, Africa and the Andean countries are being abandoned each year because severe soil erosion has made them unproductive, or at least incapable of sustaining the local inhabitants with existing agricultural technologies."

Even if we manage to add previously uncultivated land to our farm acreage, much of this new land is likely to be used just to grow food to replace the crops no longer available because other land has been lost.

There is very little reason, then, to believe that we will be able to feed the world's growing population if we insist on following traditional approaches—make more fertilizers, introduce new strains of crops, use more irrigation water, pry open what potentially cultivable land is left. If we do not recognize the limitations of our current thinking, if we do not use our imaginations to try completely new ideas, we will not be able to spare thousands of millions of people the agonies of hunger, malnutrition, and slow starvation.

5
STOKING HOTHOUSE EARTH: PART I

The prospect of thermal pollution and what it implies—superheating of the atmosphere, melting of the ice caps, swelling of the oceans, wholesale drowning of cities and countries, displacement and starvation of billions of people, the consequent upheavals of civilization—all this is in large part a prospect determined by the laws of nature, the physical principles that control the distribution and behavior of heat in the atmosphere. But the headlong momentum toward the deluge, the seemingly inexorable rush to what will be humanity's greatest act of

56

environmental destruction, is being stoked by man's mind-less drive to use ever increasing amounts of energy.

The drive to consume more and more energy, in turn, is sustained by the continued growth of man's populations. True, a good portion of this growing mass of humanity lives near, at, or below the poverty level. Nevertheless, billions today, and tens of billions tomorrow, will have the power and the desire to gather unto themselves more and more of those goods, commodities, and services thought to be the key to a higher standard of living.

As more and more people crowd the globe, every nation will have to produce and consume increasingly prodigious amounts of energy. Urban centers already in existence will grow. New urban complexes will spring up where today there are only small cities and towns. More buildings will have to be air-conditioned and heated. Tens of millions of new homes and apartment buildings will have to be illumi-nated. More and more city streets will have to be lighted. More machinery, more equipment—all of it dependent on energy and all of it discharging heat—will be put into use just to turn out the basic products necessary to maintain life. Increased populations will mean more cars and more trucks on the highway, more planes in the air, more ships on the sea: all of them transporting people from place to place, goods from factory to consumer, food from farm to city; all burning energy; all discharging heat into the atmosphere.

There is virtually no doubt, I feel, that the per capita yearly demand for energy, spurred on by growing popula-tions seeking better lives, is likely to spiral upward in dizzy-ing fashion for scores of years to come.

Estimates of the rate at which this demand will increase vary. Some experts think it will be very high. "The growth of energy . . . will stabilize around six-and-a-half to seven percent," Dr. Ali Bulent Cambel, a noted authority on energy and economic development, said in recent testimony before the Senate Committee on Interior and Insular Affairs. "I ascribe about four to five percent to increased standards of living, which all the people want, and I ascribe about two-and-a-half percent to . . . population growth." At a constant annual growth rate of 4 to 6 percent—the range used in this book—the total energy consumption of the world would double every twelve to seventeen years, and it would increase tenfold every forty to sixty years.*

The evidence of the world's growing rate of energy consumption is all around us. The use of electric power has been growing especially rapidly, and this growth is important because electrical energy, though desirable in many ways, is a wasteful form of energy because the efficiency of its production is relatively low. In 1970 the citizens of the USA used about 5×10^{15} Btus of electricity to light their streets and buildings, heat their homes, run their all-electric kitchens, activate their electric toothbrushes, drive their garbage compactors, operate their commercial firms, and power their industrial enterprises. Yet the Federal Power Commission estimates that by 1980 the USA's demand for electricity will have doubled to about 10×10^{15} Btus. By 1990, according to the Commission, the demand for electricity will have doubled once more, this time to some 20×10^{15} Btus. In fact, by the year 2000 the demand for electrical power is

---

* The calculations are straightforward. To find out how many years it will take for any constantly accelerating variable to double—a variable such as population, energy consumption, or a bank account drawing compound interest—the number 70 is divided by the annual percentage increase rate of the variable. To find out how long it will take for the variable to increase tenfold under the same conditions, the number 230 is divided by the annual percentage increase rate. The numbers 70 and 230 have been found by mathematicians to be appropriate for making these kinds of calculations.

projected to represent almost 50 percent of the total annual American consumption of energy.

If we look at the way per capita income and per capita gross national product have grown in the past few years, and if we correlate those figures with the growth in per capita energy demand, we find more evidence that our consumption of power is likely to continue escalating. Back in 1961, the average American was earning about $2200 a year and was using a bit less than 200 million Btus a year. Today, with a per capita income that is almost twice as high, the average American is consuming almost twice as much energy—350 million Btus a year.

Although we are some 5.3 percent of the world's population, we are responsible for about 35 percent of the world's total annual energy consumption. But the other nations are racing to catch up, to increase their own energy consumption at an accelerating rate. In 1961 the average Swede earned less than $2000 a year and was consuming only about 85 million Btus a year. By 1968, with his income up to almost $3400 a year, the average Swede had hiked his personal power consumption to nearly 150 million Btus. Between 1961 and 1968, the average Canadian increased his energy consumption from 125 million to nearly 250 million Btus a year. The average German increased his consumption from 75 million to almost 125 million Btus during those years.

It seems inevitable that as economies grow, as gross national products and personal incomes expand, per capita energy consumption will grow as well. There is little reason to doubt that American and Western European per capita energy consumption will reach one billion Btus per year within the next few decades. And there is little reason to doubt that as economic development moves forward, per capita energy consumption in today's lesser-developed nations will reach 350 million Btus a year—and perhaps even more.

Many experts argue that man is unlikely to continue accelerating his energy consumption at rates approaching 6 percent a year, because there is no precedent for such a sustained, furiously spiraling demand for energy. A hundred years ago, before the United States became a highly industrialized nation, it is argued, the average American was already using about 130 million Btus of energy. Today, the argument continues, with the United States leading the world in economic activity, the average American is consuming "only" 350 million Btus a year—less than three times the amount his "primitive" ancestors were using a hundred years ago. This would correspond to an average per capita consumption acceleration of only about 1 percent per year.

Superficially, this seems to be a reasonable argument. But if it is examined closely we quickly see that it is a specious bit of reasoning. If we are going to talk precisely about energy, we must distinguish between *gross* per capita consumption of energy and *useful* per capita consumption of energy.

When experts say that the average American was already using some 130 million Btus of energy a hundred years ago, they are really saying that Americans, busy cutting down the continent's forests to make way for cities, farms, factories, railroad tracks, and roads, were using the abundant wood from those fallen forests for fuel. It is true that this prodigious amount of wood carried within itself the gross energy equivalent of almost 130 million Btus per person. But that energy was used in extremely inefficient ways. Locomotives were propelled with the most rudimentary of fireboxes. Their boilers had little or no insulation, and the heat of the exhaust steam was simply thrown out into the atmosphere instead of being recycled by recondensation of the

60

vapor. Heat for many industrial activities was derived from open fireplaces. Metals were often forged out of doors over open fires. Although the gross energy consumption was relatively high, the amount of useful heat and energy derived from those primitive methods was very low. The average locomotive burning wood, the average oven in the small horseshoe factory, managed to utilize very little of the energy stored in the wood.

Thus the equivalent of about 130 million Btus per person per year was thrown into the nation's fires in the form of cordwood in the middle of the nineteenth century, but only 5 to 10 percent that much energy—perhaps 6 to 12 million Btus per person per year—came out in forms useful to man. Today, on the other hand, when we talk of 350 million Btus per person, we are really talking about 90 to 130 million *useful* Btus. We are talking about energy that is under our control, energy that is suitable for meeting our needs. We are not getting just twice as much useful energy per person as we did a hundred years ago. We are getting ten to twenty times as much useful energy as we did then.

If we wanted to carry to its logical extreme the argument of those who base their theories only on gross energy consumption and say that we are using only twice as much energy as we did a hundred years ago, we could in all seriousness say that today man is using 4000 times *less* gross energy per person than he did 30,000 years ago. At that time the population of the earth was about one million. The sun —which was man's sole source of energy because it grew the nuts and berries he consumed, grew the grass to feed the deer whose skin he wore and whose body he ate, grew the trees he burned in his little fires, and warmed his total environment—poured energy onto the surface of the earth at a rate of 2.5×10^{21} Btus per year, or about 2.5×10^{15} Btus per year for each cave man, cave woman, and cave child. Today, of course, the sun is contributing approximately the same amount of energy. And today man's activities are generating about 2 x 10^{17} Btus per year from energy sources buried in the earth. But there are close to four billion people on earth today. Thus the gross amount of energy generated by the sun, plus the gross amount of energy generated by man, divided by the total number of people, really equals an average gross consumption of energy of about 6.3 x 10^{11} Btus per person per year—4000 times *less* than when man covered himself in animal skins. Again, the point is that 30,000 years ago a lot of energy was being "consumed," almost as much as now, but it was very inefficiently used.

The energy experts who contend that there is no precedent on which to base the prediction of a sustained consumption spiral of 4 to 6 percent a year usually admit that energy consumption *will* grow in years to come, but they say this growth will automatically dwindle down to ever more modest levels. The lesser-developed nations, some of these experts argue, are not likely to increase their current per capita consumption rate dramatically because they just don't have enough economic strength to sustain energy use at a steeply accelerating pace. "The underdeveloped nations don't have the money and the resources to increase their energy consumption at such high rates," says one. "India and Pakistan for example, have had to cut back on their consumption of petroleum-based fertilizers because they cannot afford even those. The experts are unanimous that we will see growth rates no larger than 2 to 3 percent."

But I think it is unrealistic to believe that all or even most of the lesser-developed nations will fall off the energy consumption escalator just because there has been a recent rise in the cost of petroleum-based fertilizers. History makes it plain that nations will go to war or do whatever else seems necessary to improve their economic and political status vis-à-vis other nations. If an ambitious government must let many of its poorest starve today so that the nation's economic and political stand in the world is strengthened for tomorrow, the government will often proceed resolutely to do just that. "Dr. R. Ewell from New York State University recently pointed out that to just keep pace with population growth, the developing nations of the world in the next ten years will need to spend $28 billion—$2.8 billion annually—to expand fertilizer plant production and imports," Dr. Borlaug told the Senate Committee on Agriculture and Forestry. "But these nations generally say this can't be achieved. Perhaps not. But when we look at the priorities for armament we see these same developing nations collectively each year spend $26 billion on military hardware."

Some people argue that the acceleration of national energy consumption rates—especially in the lesser-developed countries—will fall off soon because the nations will quickly be coming hard up against roadblocks to finding the financial capital required for building the needed plants and mines and energy conversion systems. What such arguments ignore is the fact that the amount of a nation's money generally expands on the average in proportion to its economy as a whole, that the growth of a nation's capital availability tends to accelerate right along with its population and energy consumption rate. Russia, for example, was flat broke in 1917, but now, less than six decades later, she stands forth as a global colossus; somehow she created the necessary capital to build up her strength. The People's Republic of China appears to be following a similar track. A nation's will power and organizing ability are more important than its quantities of foreign exchange, I believe, in determining how

63

rapidly it can accelerate its growth rate, even if it is a lesser-developed country.

As for more developed countries, some experts argue that they have reached plateaus in their consumption of energy because there just isn't very much more the citizens of those wealthy countries could do with more energy. "After a while you can only use so much energy no matter how much you toot around in your third car or your fifth motorboat," says one of the specialists. "After all, you can only use one vehicle at a time."

The trouble with that argument is that it ignores the energy required to *make* all the cars and motorboats, regardless of how much they are used. I estimate that some 20 to 25 percent of the total energy consumption an automobile represents is used in building the car plus the factories, garages, sales rooms, and so on, for delivering it to the customer, and only about 75 to 80 percent of the total goes into operating the car during the years it is used. The argument also ignores the energy consumed by all the collective enterprises of the world—the armies, navies, and air forces, the space projects, and other vast energy consumers—which no single individual ever "uses" at all.

I am doubtful that any of the "growth will automatically stop" arguments will stand the test of time. Historically, experts seem to have consistently underestimated man's desire and power to grow. At first glance it seems true that the developed countries already enjoy a surfeit of energy consumption. But even if families in the developed countries stay small, even if the ideal of zero population growth in the industrialized nations is attained, energy demands, I am convinced, will nevertheless tend to surge forward.

The argument that countries like the United States are already using all the power they possibly can ignores the fact that even in countries like America there are vast numbers of people who are still living on low incomes, at or near the poverty line. When we say that the average American uses 350 million Btus of energy a year, we are talking about a mythical "average person." Millions of Americans do enjoy a life style that consumes more than 350 million Btus each year, but there are many more millions of Americans who have to get along on considerably less. The people who are now disadvantaged—who have to walk to the nearby welfare kitchen while their neighbors drive to distant but rewarding jobs—will naturally be trying to catch up. At the same time the rich will be trying to stay ahead. We must therefore conclude that the average Btu consumption in this country is going to rise, and rise considerably.

Even if we were to assume that in time every single American will be enjoying an upper-middle-class standard of living, we would still have no reason to believe that energy consumption will level off. It is true that nobody can drive more than one car at a time. But powerful people have a curious way of needing and seeking and acquiring and using more and more power. People who are already using a good deal of energy driving their automobiles on weekdays, racing their power boats on weekends, dashing about on business or vacation trips in 747s, tend to develop more and more expansive ambitions, some of which can be satisfied only with proportionate increases in their already large energy consumption rates.

If, in 1850, a time-displaced pollster had stopped a long-skirted lady on the street to ask whether or not she needed something called an automobile, the lady probably would have reacted by backing away very slowly until she could make a dash for safety. Even if our nineteenth-century lady had stuck around long enough to be told what an automobile is, chances are good that she would have asked, "Well, what on earth would I need it *for?* I'm doing very well just as I am, thank you!"

If the same pollster had popped up in the late 1890s or early 1900s, to ask people who were just getting used to the horseless carriage whether they could foresee a need for Chevrolets, Plymouths, Corvettes, Mercedes-Benzes, Pontiacs, Hondas, Peugeots, Toyotas, Volkswagens, Cadillacs, Datsuns, Jaguars, Fiats, Lincoln Continentals, Porsches, Chrysler Imperials, station wagons, two-door sedans, four-door sedans, piston-powered cars, rotary-engine cars, air-conditioned cars, cars with stereophonic equipment, with push-button windows, with power brakes, power steering, power door locks, dune buggies, snowmobiles, campers capable of sleeping two to eight people—the initial reactions would have been the same: What on earth do we need all that *for?*

But when someone invents something attractive, it remains a simple luxury only a short time. Soon it becomes a commodity, then a need, then a need with a thousand and one variations. Ovens become double ovens, self-cleaning ovens, microwave ovens. Refrigerators become self-defrosting refrigerators, ice-cube-generating refrigerators, refrigerators bejeweled with AM-FM radios. Garbage cans become garbage shredders and then garbage compactors. As soon as anything becomes an established part of our consciousness, no matter how expensive, no matter how hungry for energy it is, great new industries arise—industries in themselves great consumers of energy—to mass produce that product in countless variations.

Why assume, then—and thereby fly in the face of historical precedent—that we have reached the peak of our energy-consuming wants? How do we know what magical things are still to come? Will we turn down the self-cleaning house just because the dust sensors and automatic sweepers use a hundred times as much electricity as present-day vacuum cleaners?

We cannot escape the conclusion that competition among nations, like competition among individuals, leads countries and people to follow paths that are not necessarily in their best long-range interests. If one's status on a day-by-day basis is obviously inferior to a neighbor's, one feels a very powerful and continuing spur to increase one's consumption rates. This is not just a matter of personal or national pride, though that is often of great importance; it is also, very often, a matter of immediate survival. On the other hand, to perceive and to act on global problems or the long-range strategic interests of mankind requires a difficult feat of the educated imagination.

Therefore, no matter how much we would like to believe otherwise, we must recognize that this intense competitive drive, this need we all feel to enrich ourselves as individuals and nations, is driving us to devour more and more fossil fuels, to develop and use more esoteric forms of energy (including nuclear fuels), to heat up our atmosphere and intensify our attack on the ice caps, and is propelling us into catastrophe.

6
STOKING HOTHOUSE EARTH: PART 2

The years to come, we are warned, will bring more and more "brownouts," "blackouts," cold and heatless winters, hot and nonair-conditioned summers, restricted food supplies, curtailed driving, and a host of other personal and national hardships brought on by a severe shortage of energy.

Can we reconcile these warnings with visions of major metropolitan and agricultural areas under water, of rolling, white-capped waves lapping at the lower windows of the Empire State Building, the glass and aluminum high rises in London, the statues in Copenhagen, the Colosseum in Rome? Given all the talk about rapidly vanishing natural resources, can we believe that we will create, by our enormous production and consumption of power, a superhothouse atmosphere that will melt the polar ice and thereby drown and devastate much of the civilization we have built over thousands of years?

I think we can. To a great extent we are suffering an energy crisis today for reasons that have nothing to do with real deficits in fuel supplies. Governments of nations that import fossil fuels to sustain their economies have seldom bothered to study carefully the role of these fuels in their countries' well-being. They have apparently assumed that cheap and plentiful oil could always be bought from subservient sheiks. The governments never bothered to note that the sons of those sheiks were studying at Harvard Business School and the London School of Economics, learning every modern way to monopolize production, raise prices, cut supplies, and play consumer against consumer. Nations in which most of the fossil fuels are now extracted have learned to use oil to achieve their own political objectives. Fuel flows are distorted by political tugs of war between Arabs, Jews, Americans, Russians, Europeans, and Japanese. And naturally the big oil companies are doing their best to turn big problems into big profits. All these factors have combined to produce shortages at the consumer's level. There are no real shortages as yet at the producer's level: there is more than enough oil out there to meet our current needs.

I think, too, that the energy deficits predicted for tomorrow—that is, for the next ten to twenty-five years—are predicted largely by people who make it their business to be extremely conservative in their estimates of the fuel supplies remaining beneath the earth.

Many of those whose opinions are sought about impending energy supply shortages are scientists whose professional and personal standards of honesty apparently require them to state only what they "certainly know" rather than "probably know" or "reasonably suspect"; this inevitably produces severe underestimates of the amounts of resources remaining to be discovered and pinpointed. Many of these experts are executives and scientists employed by companies and industries that extract and refine the fossil fuels on which we depend today; they often appear to have a vested interest in underestimating the amount of oil, coal, or natural gas beneath the ground.

Ultraconservative estimates may help to keep down the tax burdens of major fuel corporations. If a company has bought or controls a vast amount of land because there might be oil, coal, or natural gas underneath, the company pays minimal taxes—just on the valuation of the topside land —as long as no one has proved that there are indeed minerals there, and as long as there are no estimates of actual amounts beneath the surface. When a dollars-and-cents estimate is finally made of the fossil fuel that is probably down there, the company's tax bills might tend to shoot up, so it is understandable that executives might feel that they are better off not knowing—and most especially not having anyone else know—how much fuel may really exist. One thing is certain: information about all the resources controlled by an oil company is kept highly confidential within a small circle of executives in the company until it becomes absolutely necessary to release it.

Conservative estimates have also been encouraged by depletion allowances written into our tax codes. A businessman who buys machinery for his factory is allowed to cushion the impact of the financial expenditures by deducting a fraction of the price of the new machinery from his gross annual

70

income before figuring his taxes. If, for example, a piece of equipment costs $250,000 and is expected to have a useful life of ten years, every year for ten years the businessman can deduct one-tenth of that $250,000 from his gross annual income. This lowers his net income, on which he pays taxes, and thereby saves him some tax money.

Years ago the oil companies convinced the government that fossil fuels should be treated like machinery or other capital equipment—in other words, that crude petroleum is really a part of the firm's capital. Just as a piece of machinery depreciates over time, the oil companies argued, so the supplies of raw materials in the ground decrease as they are extracted. In effect, a high depreciation allowance would be a compensation to the firms for eventually having to go out of business. (Presumably the arguments were made before corporate diversification became the vogue.)

Government officials bought this argument, and the oil industry was given what is known as a percentage depletion allowance. For many years the individual companies have been allowed to reduce their gross income by some substantial percentage before beginning to calculate their tax debts to the government.

It is argued—and perhaps to some extent rightly so—that the depletion allowance actually spurs exploration for new sources of fuel because the tax privilege mitigates the high cost of the search. But the fact remains that the depletion allowance has worked to keep estimates of remaining oil supplies low. The smaller the assessment of oil remaining under the ground, the stronger becomes the theoretical argument that high depletion allowances are necessary, because the day is thus brought nearer when the company must close its doors or look for new sources of income. Small wonder, then, that fuel circles are dominated by people who will not admit to the presence of fuel unless they can bend over and pick it up, see it, smell it, or taste it.

Even if we conclude that energy consumption levels will probably rise for many years to come, it is clear that there are more than enough raw materials beneath our feet to sustain our hot pursuit of energy long enough to bring us well into the period when man's emissions of heat into the atmosphere begin to mount a significant assault on the world's ice caps.

In 1970, the estimated proved U.S. reserves of crude oil, according to the oil industry's National Petroleum Council, were 425,200,000,000 (about 4.3×10^{11}) barrels. It has also been estimated by numerous experts that the world contains some 4.8×10^{12} barrels of oil—an elevenfold higher figure— still hidden and waiting to be discovered. Although some of this oil is underneath land, most of it probably lies beneath the ocean, under the offshore continental shelves and slopes and deep bottoms. Geologists believe that areas such as the North Sea, the Persian Gulf, the Irish Sea, Hudson Bay, the Caspian Sea, the Red Sea, and the Adriatic may be vast vaults holding substantial amounts of oil.

While some experts (in and out of the oil industry) talk about billions of barrels of oil yet to be discovered, there are also vast amounts of oil that have already been discovered but remain untapped. We have not yet begun to consume the oil reserves of the Arctic Region. The Russians, the Mexicans, the Norwegians, the English have yet to begin work in earnest tapping the known oil reserves in areas under their jurisdiction. China, which reportedly is sitting on oil supplies as great as those of the Middle East, has only lately begun to show an interest in its holdings.

Moreover, at least 60 to 80 percent of the original oil is still left in oil wells that have been or are about to be shut down. As long as crude petroleum was selling for two to three dollars a barrel, and as long as large depletion allowances prevailed, the oil companies were extremely wasteful in their exploitation of existing oil fields. It was mostly the oil that came up out of the wells under the force of natural subterranean pressures—pressures generated by the presence of natural gas, for example—that was hauled off to market. Oil that remained in the wells after the pressures had subsided because the natural gas had been consumed or dissipated, and oil that remained because it was too thick to flow up easily into the waiting arms of the oil companies, was simply left behind. According to most experts, the amount of crude petroleum left in old, abandoned wells equals a good 70 percent of the original volume of oil in those wells.

The oil companies have, until now, argued that this oil was "not recoverable"—or, if pressed, that it was "not economically recoverable," that it was much too expensive to buy and install the machinery and equipment necessary to bring up the harder-to-extract petroleum. Indeed it was, given the fact that those companies faced a cutthroat competition to stay in business. But now that oil is increasingly dominated by world-wide cartels which sell only at prices much higher than those of a couple of years ago, oil companies are taking a second look at the oil fields they jilted so quickly. Many firms are experimenting with special machinery that can inject hot steam or other forms of heat into old oil reservoirs to liquefy overthick petroleum and make it flow easier. Others are buying equipment that can fill old oil wells with new sources of pressure and force them to relinquish the petroleum that is still there.

Continental Oil engineers, for example, have pumped special detergents under high pressure into a few experimental wells. The detergents are expected to seep through the thick oil remaining in the wells, thinning it out sufficiently to allow it to flow to the surface. If the experiment works, Continental Oil officials say, the procedure could bring up as many as two million barrels of oil a day, as much as the production of oil from the much-vaunted North Slope of Alaska. Many oil experts now feel that the new efforts to recoup abandoned oil will be successful, and that the overall productivity of existing oil wells can be increased from the existing value of about 20 to 30 percent to about 50 percent. If so, the proved crude oil reserves in the United States alone would jump 80 billion barrels—as much petroleum as has been produced in the United States since the first well was drilled in 1859.

The huge petroleum reserves—those already discovered and tapped, those discovered and not yet mined, and those still to be found—were formed during a vast stretch of time that began some hundreds of millions of years ago. (Indeed, the process is presumably still going on, and at about the same rate as in primordial times. But the natural process is excruciatingly slow. According to some geologists, it will take nearly another billion years for that process to generate the amount of oil created up to now.)

During the early part of the earth's geological history, some small fraction of the living things—notably single-celled plants and animals that lived in the oceans, rivers, and lakes, as well as the more complex plants and animals that flourished on land and in swamps—did not decay immediately when they died. As the single-celled plants and animals died, they sank to the bottom of the watery environments where they lived. The larger dead plants and animals sank into the swamps where many of them had lived and flourished, or were ultimately swept from dry land into the rivers and seas. Over millions of years, erosive processes deposited layer after layer of soil and mud over the dead organisms

74

that had collected in various areas. Covered by this earth (which under increasing pressures eventually turned to rock), the masses of organic material slowly turned into great subterranean pools of oil, the pools that we tap today with drills.

Not all the organic material that died was conveniently turned into oil and collected into great reservoirs. Much oil, tar, and solid organic material is found to be thoroughly dispersed through sands and strata of rock called shale. These deposits of organic matter may represent billions and billions of additional Btus of energy to meet our future fuel demands.

Because the organic material present within the shale is scattered throughout the body of the rock, experts have had to think up new technologies to extract it. In one process, the rocks containing the organic material are mined and transported to a refinery. There they are placed in furnaces where fires ranging from 800 to 1200 degrees Fahrenheit force the organics to vaporize. The vapors are then converted into a thick oil, and the oil in turn is processed into various products, including fertilizer, gasoline, and plastics.

The process in full swing could chew up some millions of tons of rocks a day, an operation that would leave ugly scars in the earth in the mining area as well as ungainly mountains of waste rock at the processing site unless huge land reclamation efforts were required by law. Such law is being sought by environmental activists, and meanwhile the oil industry is experimenting with technology that would allow the extraction of oil from shale by vaporizing the organic material right at its location deep in the earth. This on-site extraction would start with a hot fire ignited in a deep well drilled into the shale rock. The fire would melt and vaporize the individual bits of organic detritus out of the rock surrounding the well, and the resulting oil and vapor would then flow into other wells dug around the periphery of the fire well. The oil and vapor would then be pumped up through these wells to the surface for direct transport to a refinery.

There is, of course, a good deal of speculation and debate about what contribution shale oil will make to our energy needs in the years to come. Some experts denigrate shale as a potential source of oil. The pessimists say that the world's shale oil will last for only a few years after we have begun full-scale extraction, because by the time we have perfected the necessary technology (in the next twenty-five years) we will be consuming oil at a much faster rate than we are today. They also say we will not be able to squeeze more than about 1300 billion barrels of oil (about 10^{19} Btus) out of the shale.

Other experts however, doubt these gloomy appraisals. They say that the world's total reserves of shale—reserves speculated upon but not yet conclusively proved to exist— could total 10^{22} Btus, an amount a thousand times larger than the pessimists give them credit for.

Although we talk a great deal about oil and its role in our energy future, it is not the only fossil fuel that will help us melt the ice caps. Coal, the fossil fuel we depended on long before oil became the leading supporter of man's industriali- zation, is still here in plenty. Some of that coal will be used to raise steam to run the turbines producing electricity, but much of it can and will be used as a raw material for the manufacture of other fuels, including oil, gasoline, and syn- thetic natural gas. There are ample precedents for this use of coal. During World War II, when Germany was cut off from her traditional sources of petroleum, German scientists be- came highly skilled in the art of converting coal to gasoline. By 1944 Germany was manufacturing gasoline from coal at a rate of more than 200 million barrels per year for the

German war machine. In this country industrial scientists are using coal to make about one billion gallons of methanol (methyl alcohol—"wood alcohol") every year. Although methanol currently costs quite a bit more per gallon than gasoline does, if production were increased the price could drop sufficiently to make methanol supplemental to or competitive with gasoline.

If present-day worries about energy crises and energy shortages seem to be incompatible with my forecast that we may soon be using enough energy to upset the delicate atmospheric heat balance sustaining the ice caps, we should also remember that, fossil fuels or no fossil fuels, scientists and technologists are pressing forward to furnish us with new sources of energy.

Although the earth's surface is cool enough to allow us to walk around on it, the lower depths of our globe are composed in large part of hot or molten rock. In some places, this molten rock, called magma, seeps upward toward the surface. It does not usually flow straight to the top, but it does reach high enough in many places to pass into fractures in the rocks that make up the earth's outer layer, the crust. When magma reaches some of these fractures, its great stores of heat are simply radiated, convected, and conducted out into the atmosphere. When the magma oozes up through other crustal openings, however, it meets submerged water supplies—underground streams and water tables—and turns them into steam that escapes to the surface under great pressure. The process is somewhat like the one that takes place on the kitchen stove in a covered kettle of water. As heat is supplied to the water, both the water temperature and the pressure of the vapor (steam) under the lid rise. As the temperature rises more and more, the vapor builds up to a higher and higher pressure in the kettle, until finally it reaches such a value that it can lift the heavy lid and escape. In some places around the globe, the boiling of the subterranean water is continual and vapor constantly escapes into the air through the fractures in the crust. In other places the heating is constant but the boiling comes intermittently, and the escape of the vapor up through the fractures is therefore intermittent too. Probably the best-known example of intermittent vaporization and escape is Old Faithful, in Yellowstone National Park. Every time this geyser (or any other geyser like it) blows sky high, it is a signal that a water source within the earth has been heated until it has achieved sufficient vapor pressure to enable it to escape its underground containment.

The magma's heat itself, and the magma-produced steam and vapor escaping the bowels of the earth under great pressure, are obvious sources of energy. This energy flows up into the atmosphere, and it does this under present conditions at about three or four hundredths of 1 percent of the rate at which solar energy is received at the surface. In many parts of the world, this natural source of heat energy— geothermal energy—is already being used in a modest way to fulfill some topside energy needs. Geothermal energy fields have been driving electric turbines and heating buildings for some time in Italy, Iceland, New Zealand, and Japan.

Many economic and technical questions have to be solved, however, if geothermal energy is to become a major energy source. If vaporized water were to be put into pipelines for shipment to a power plant, it would, unless expensive insulation were installed around the walls of the pipes carrying the water, lose much of its heat energy before it arrived at the turbines waiting for it. Because of this, most geothermal experts feel that American utility companies or fuel companies that want to take advantage of geothermal fields must build new power plants right at the geothermal site. Since this would involve huge capital outlays, long, difficult, and costly studies would have to be carried out to prove that a particular geothermal field was likely to last at least thirty years, the amount of time over which a plant is depreciated against taxable income.

Vaporized water that shoots up out of the earth is often laden with corrosive minerals and salts that could hamper or destroy equipment used to turn escaping steam and water into electricity. Water emerging in geothermal fields near California's Salton Sea, for example, has a 20-percent salt content—about six times as much as there is in conventional sea water.

Scientists are sure, however, that many of the technical problems can be worked out. Water full of minerals or salts could be processed and demineralized before it reaches the turbines. It could then be used for other purposes, including irrigation of nearby agricultural lands. As an alternative, the water—minerals and all—could be passed through a series of heat exchangers that would use the geothermal energy to boil another, more neutral fluid, which would then be used to drive the turbines.

Many scientists even believe that there is no technical reason why geothermal fields could not be developed where none now exist, because hot rock can be found at sufficient depths at any point under the earth's surface. Artificial cracks and openings deep in the earth could be produced with hydraulic pressure or dynamite (or controlled nuclear explosions), and water from the surface of the earth could then be pressurized and injected through wells drilled into these zones of artificial fissures. Once the water had been heated by the hot rock, it would return to the surface through

79

other wells and would be ready for use in the production of electricity.

Confident that geothermal energy will become an important source of power, scientists and technicians around the world are intensifying their efforts (including the use of infrared photography from space satellites) to find new geothermal fields and exploit existing ones. Japan's known geothermal fields will probably yield huge quantities of electric power within a short time. Just north of San Francisco, Union Oil is already producing and selling hundreds of megawatts of power from a geothermal field known as The Geysers. It is expected that by 1981 the field will be producing over 1000 megawatts of electric power. Federal officials have estimated that geothermal fields in the United States could be providing almost 20 percent of the country's energy requirements by the turn of the century. Thirty other countries—including Chile, Ethiopia, the French West Indies, Indonesia, Kenya, Nicaragua, the Philippines, and Turkey—are also exploring potential geothermal sites. Ultimately, some experts think, geothermal energy could be giving the world more than 100,000 megawatts of power (3×10^{15} Btus) per year.

Of course, one thing we can be sure of: any greatly intensified use of geothermal energy would contribute to the world's thermal pollution problems.

Despite the increased exploitation of the earth's fossil fuels and its internal heat supplies, most energy experts (at least until quite recently) have been putting their bets on the budding nuclear industry. It is the nuclear reactor, they think, that will supply man with most (maybe all) of his energy needs after the turn of the twenty-first century.

Many people look upon nuclear energy as a gift from the gods because, using a relatively small amount of raw mate-

80

rial, it can generate vast amounts of energy. A conventional power plant that produces 1000 megawatts of electric power, the nuclear enthusiasts point out, consumes over two million tons of fossil fuels a year, but a 1000-megawatt plant powered by a nuclear reactor uses only thirty-five tons or so of uranium a year.

Nuclear power advocates are optimistic about nuclear technology yet to be perfected. Conventional reactors now in use across the country produce heat by splitting, or fissioning, the U-235 nuclei contained in the uranium. Natural uranium contains atoms of several kinds—about 0.7 of 1 percent of the atoms are U-235 and the rest are mainly U-238. "Reactor-grade uranium" (processed so as to contain almost no impurities capable of absorbing neutrons, and also to contain a somewhat higher than natural concentration of the valuable U-235 atoms) is inserted into the core of the reactor in thousands of tubes, or fuel rods, about 12 feet long and about as thick as a lead pencil. These tubes—or pins, as they are more commonly called—are bombarded by slow neutrons. As the neutrons strike the pins, they split the U-235 atoms. The heat given off by the U-235 as a result of the fissioning, or splitting, process is carried away by water that is constantly bathing the outer walls of the fuel rods. The water is then allowed to vaporize into steam for driving turbine generators. There are more than four dozen such commercial nuclear plants operating in the United States already, and present government plans call for a total of 245 to be operating and turning out approximately one-third of the USA's electrical energy by 1985.

Despite the great amount of energy they produce, today's nuclear reactors are basically inefficient. A pound of natural uranium—a nugget no bigger than a good-sized walnut—contains as much energy in its U-235 nuclei as all the coal

81

that could be piled into a string of freight cars. But today's conventional nuclear reactor taps less than 1 percent of the total energy available in the pound of uranium, because the reactor fissions only the U-235 in the mineral. In these "simple" nuclear reactors the U-238 is wasted because it will not split nor give off energy under the impact of slow neutrons.

In more advanced, "fast-breeder" nuclear reactors, raw uranium can be used much more efficiently and therefore can be made to yield truly prodigious amounts of energy per pound of fuel. Reactor-grade uranium with its enriched load of U-235 is placed in the pins within the core of the reactor, and additional pins containing "depleted uranium" (just U-238 alone) are placed around the core. This type of reactor, which is still in the early stages of its development, uses slow neutrons to fission the U-235 within the core, thereby producing power. But then the fast neutrons, which are given off by the fissioning U-235, are absorbed by the U-238 in the core and in the blanket of pins around the core. The result is not just another nuclear reaction but the "breeding" of a new reactor fuel—the transformation of large amounts of U-238 into plutonium 239, a material that is exceedingly rare in nature. After the plutonium 239 has been extracted from the U-238 by special chemical techniques, it too can be used as fissionable material because it will allow itself to be split by slow neutrons just as U-235 does. This breeding process can be repeated again and again, until the original amount of raw uranium has been forced to give up forty to eighty times more energy than it would have given up in the conventional nuclear reactors in use today.

If breeder reactors were used to process all of the world's potentially fissionable fuels (including thorium, which can be converted into a fissionable fuel by immersion in the intense flow of neutrons in the interior of a nuclear reactor), the result would be the release of more than a thousand times as much energy as could be obtained by burning the world's entire resources of all types of fossil fuels.

Many experts are convinced that technological advances will take us beyond the breeder reactor to an even more powerful source of nuclear energy—fusion. Today's reactors create energy by splitting—fissioning—heavy nuclei; the fusion reactors of the future may produce energy by welding light nuclei together. The fusion of hydrogen nuclei is responsible for the production of energy in stars like the sun, and it is the fusion of light nuclei that makes possible powerful weapons like the hydrogen bomb.

Although scientists can design and build hydrogen bombs, they do not yet have the very sophisticated technical ability that would enable them to carry out controlled fusion within a small power plant. It is one thing to unleash a massive force out on some atoll in the Pacific Ocean but quite another to try to control a similar force operating within the confines of a nuclear reactor. To control fusion, scientists will have to heat an electrically charged gas, called plasma, to temperatures much higher than those in the interior of the sun, at least to about 200 million degrees Fahrenheit. Moreover, they will have to sustain these temperatures for at least a few hundredths or even a few tenths of a second—and sustain them without allowing the superheated gas to touch and disintegrate the walls of the container. Finally, they will have to design a reactor that can carry out this exacting maneuver continuously or repetitively to generate a continuing stream of power.

Experts differ on the fusion reactor's possible date of arrival. Some say that the obstacles blocking the road to controlled fusion are so great that practical reactors are still many decades away. Others maintain that the fusion reactor is much closer. Whatever the disagreements on time, most people connected with nuclear power are anxious to get on with the job of developing the fusion reactor because they believe it will give man a virtually inexhaustible and relatively clean supply of fuel to fill even his most ambitious energy demands. Fusion reactors will most likely be able to use a heavy form of hydrogen called deuterium, and this substance is abundant in the waters of the earth. The deuterium in one gallon of water has the energy equivalent of more than 300 gallons of gasoline. In fact, experts estimate that there is enough deuterium available for fusion to give mankind some thousandfold more Btus than are contained in all of the world's potential sources of fissionable fuel, or a millionfold more energy than is present in the world's total estimated potential resources of fossil fuels.

Over the past few years we have become painfully aware that our use of fossil fuels has diminished the quality of the environment in which we developed and thrived. Although the laws of nature tell us that thermal pollution is the only *necessary* and *unavoidable* form of pollution which a greatly increased energy use will inflict upon us, the "natural laws of human nature and human societies" make it appear rather certain that any intense effort in the next seventy-five to one hundred twenty-five years to step up the use of the fossil and nuclear fuels available to us will increasingly subject us to *unnecessary* and *avoidable* pollutions of a nonthermal kind. The search for oil will almost certainly bring more explorations and operations in delicate environments like the Alaskan tundra, more interference with the fragile ecological systems that have evolved in regions previously undisturbed by man. Increased and intensified use of coal will bring more mines; hence more miners will probably be crippled or killed by lung-choking dusts, exploding methane gas, or other dangers. If strip mining may proceed without being obliged to restore the ravaged countryside, then it will expand rapidly, leaving in its wake more destroyed landscapes, more acid or silt-filled streams, more abandoned communities.

Intensified use of fossil fuels will almost certainly mean more air pollution too, especially if coal comes to be used in increasing amounts to produce power after oil really starts to run low. Smoke is relatively easy and cheap to filter out of the stack gases of coal-fired power plants, but separating sulfur from coal has proved to be a somewhat more expensive problem. Some companies have tried to develop equipment that will remove sulfur from the gases that find their way into the stacks of power stations after the coal is burned, but these and other efforts have not yet gained wide acceptance within the power industry.

A general feature of pollution control or pollution cleanup activities, of course, is that neither plant managers nor company presidents can introduce them "voluntarily," no matter how socially desirable they seem to be, if the effect of introducing them is to reduce the immediate profits produced by the plant or company. This simple truth follows from the fact that if such "voluntary" actions *were* taken by a responsible company official, then he would immediately be fired from his position or else his plant or company would be likely to go out of business more or less quickly in the face of competition from less scrupulous rivals.

Therefore, the only way that socially desirable cleanup activities can be introduced is by using the law to penalize— by fines or taxes or criminal penalties—all who fail to introduce the desired activities as required. When the law is used this way, the effect is to shift the "rules of the competition." The newly required cleanup expenses become an accepted cost of doing business for all alike, and the socially desirable activities are introduced reasonably rapidly.

Even geothermal energy—ostensibly a "clean" form of

energy—can create pollution. The same wells that produce the high-pressure steam that would be used to drive turbines also produce a host of noxious gases and liquid discharges. Some can be sorted out and contained before they find their way out of the stacks into the air or out of the sewer into the water supply. But one gas—hydrogen sulfide, a form of sulfur—has proved to be expensive to control. As a result, some energy experts think that sulfur emissions from geo-thermal plants might be as difficult a problem to solve as that of sulfur emissions from power plants using fossil fuels.

The same kinds of hard environmental questions have to be asked about our headlong rush into nuclear power development.

Fission plants produce an immensely dangerous type of waste. These wastes, coming from the very depths of the nuclear reactor, are intensely radioactive. They will have to be carefully stored out of the way of us human beings, our cities, our farm lands, and our water reservoirs, for thousands of years, until the wastes have "cooled down" enough to be safe.

But where to store all this material? Even now, when atomic energy is producing only a very small percentage of the nation's power, the United States government already has some 90 million gallons of nuclear waste (produced largely as a result of the creation of nuclear weapons) on its hands. No one really knows what to do with these wastes or with future nuclear residues. Some have suggested that nuclear wastes might be placed on rocket ships and driven out into space or into the sun itself. But what if the rocket ship were to blow up on the launching pad? Or, once off the pad, suppose it blew up before it managed to escape the earth's gravitational pull? It is not hard to imagine that such an accident could contaminate large areas of the world.

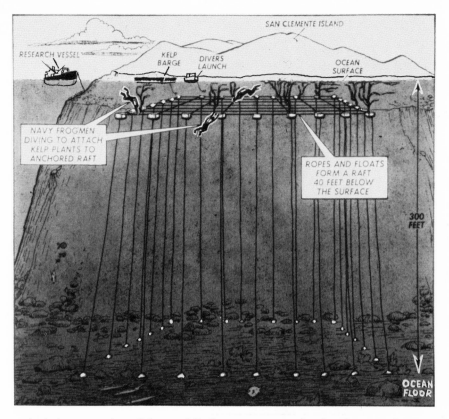

Artist's conception of the world's first open-ocean food and energy farm. The farm was established in the Pacific Ocean near San Clemente Island, sixty miles off the coast of California. *(Drawing by Russell Arasmith for the* Los Angeles Times)

Early 1974: Navy divers prepare to install and submerge the first open-ocean farm.

(U.S. Navy Photo)

The grid lines
in the foreground,
made of plastic
(polypropylene)
fibers, are ready
to be submerged
for ocean farm.

The ocean farm, with
a buoy in place, being
pulled off the Navy
ship into the sea.

Divers attach a giant kelp plant to the grid of the ocean farm.

(U.S. Navy Photo)

A commercial kelp-harvesting ship cuts the canopy of a giant kelp bed off the coast of Southern California.

(Kelco Co. Photo)

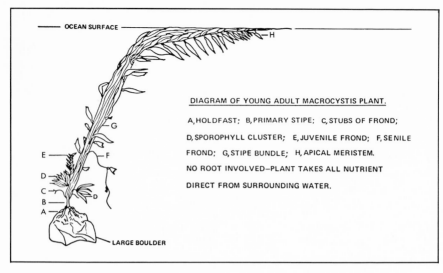

Diagram of a young adult kelp plant.

(Courtesy of W. J. North, California Institute of Technology)

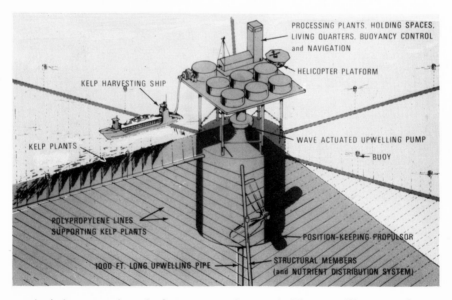

Artist's conception of a future ocean farm unit. The upwelling pump is powered by waves. The farm can be positioned by anchoring the central platform or by propelling the platform with wind power, wave power, or fuel derived from the seaweed harvest. The depth of the farm's mesh may be increased during large storms to reduce the risk of damage.

(U.S. Navy Photo)

Others have proposed that the nuclear wastes be stored in abandoned salt mines, but many fear that local water reservoirs or other valuable resources might later become contaminated. I would suggest that the best place to store them would be in artificial, cement-plugged caverns created at the bottoms of holes drilled very deep down—five miles or more —under the surface of the earth. But every method has its difficulties. Though earthquakes and other natural hazards would not seem to constitute any particular danger for deeply buried radioactive waste containers, the containers would be vulnerable on their way from the processing plant to the storage site, and the waste processing plant itself is a focal point for extremely serious dangers. There is simply no adequately safe way to take care of this immensely lethal radioactivity.

Government officials as well as many scientists worry that in these days of urban guerrillas and other fanatics it might prove all too easy for some dedicated band to break into a nuclear power plant and threaten to blow it up, or to steal enough radioactive material to build a homemade atom bomb. Theodore B. Taylor, a nuclear power expert who used to design atom bombs at Los Alamos, New Mexico, recently told a congressional subcommittee that "under conceivable circumstances a few persons, perhaps one person working alone, who possessed about ten kilograms of plutonium . . . could, within several weeks, . . . design and build a crude transportable fission bomb . . . very likely to explode, with a yield equivalent to at least 100 tons of high explosive, and that could be carried in an automobile."

Although nuclear terrorism is a very serious possibility, I worry equally about something the government doesn't seem to care enough about: accidental leaks of radioactive material from nuclear power plants.

If these nuclear plants were ordinary concrete or steel buildings, neutrons from reactions occurring in the core of the reactor would spray out through these conventional materials and contaminate the surrounding countryside. To prevent this, very thick shields are placed around the reactors. The stray neutrons produced during nuclear reactions are absorbed mainly by the shield, and as a result, direct outside radioactivity is pretty much limited to this protective layer. Theoretically, these shields should make the atomic plants safe and secure. But in practice, they don't. All too often gases and liquids used in the plant become highly radioactive and manage to leak to the outside world. For example, on December 12, 1952, an accident occurred (as a result of a "human error") in a nuclear reactor at Chalk River, Canada. This accident, once triggered, avalanched automatically through a lengthy series of "steps" occupying some seventy seconds of time, and it culminated in several high-pressure steam explosions plus a hydrogen-air explosion which completely destroyed the reactor core and produced uncontrolled releases of radioactive materials at various locations in the area. In 1957 another and more serious accident in northwest England blew large amounts of radioactivity into the air over England and northern Europe. In 1961 a reactor exploded at an isolated spot in Idaho—an explosion which killed the reactor's three operators and hurled radioactivity far and wide into the surroundings.

As the nuclear industry proliferates, such "accidental discharges" of radioactivity can be expected to increase in frequency. Indeed, as this book is being written I find in the daily newspapers the following items:

Jan. 30, 1975—"The discovery of cracks in the emergency cooling system of an atomic reactor in Illinois has forced the

Government to order utilities operating half of the nation's reactors to shut down within the next 20 days and search for similar possible faults in their power plants.

"The order marked the second time in four months that most of the same utilities had been ordered to inspect for possible cracks."

—New York Times, p. 1

March 18, 1975—"San Onofre is potentially a good place to build 'a whole string of nuclear power plants,' a noted scientific advocate of atomic energy said yesterday.

"Dr. Alvin M. Weinberg . . . said he is convinced that the danger from dams and coal burning is greater [than from nuclear reactors]."

—San Diego Union, p. B-4

March 24, 1975—"The San Onofre Nuclear Power Plant permit is in question because of a potential radiation hazard which could exclude the public from part of the state beach."

—San Diego Union, p. B-4

March 24, 1975—"The Nuclear Regulatory Commission said yesterday it had asked nine atomic power plants to quickly check their safeguards against accidents such as the fire that shut down two reactors at an Alabama plant Saturday."

—San Diego Union, p. A-13

March 26, 1975—"Early last Saturday . . . an electrician . . . held a lighted candle near some insulation to find out whether air was leaking . . . under the control room of T.V.A.'s two giant nuclear reactors.

"Seven hours later, a fire touched off by the candle had brought the reactors . . . to a halt. . . .

"Jack R. Calhoun, chief of the T.V.A.'s nuclear generation branch, confirmed . . . that the emergency core cooling system

89

–the prime system that would be called into action should there be a large loss of the regular coolant–had not functioned because the control cables had been destroyed."

–New York Times, p. 7

March 29, 1975–"A few pounds of plutonium suffice to make an atom bomb. Ingested into a human body, even a minute amount of plutonium can be lethal, for it is the most poisonous existing substance. These facts are well known.

"Yet incredibly enough two shipments of plutonium, each weighing about 100 lbs., were recently flown into Kennedy Airport, with full official permission. The shipments were allowed even though the plutonium was packed in containers that would have broken into many pieces had the plane crashed."

–New York Times, p. 22

March 30, 1975–"Radiation that leaked into a nuclear power plant heating system . . . went undetected because a monitor . . . was not operating. . . .

"Northeast Utilities . . . said there was no harm done, but gave no explanation why the detector was not working. . . .

"The contamination was discovered when a tank . . . over-flowed onto the boiler room floor. . . ."

–San Diego Union, p. A-5

April 2, 1975–"Northeast Utilities said [a second leak of] 2,500 gallons of [radioactive] water 'inadvertently' was re-leased into . . . Long Island Sound.

"No workers were overexposed to radioactivity in either leak. . . ."

–San Diego Union, p. A-10

This list is a small sample. Truth to tell, the U.S. government has not seemed to take these accidental explosions and leaks and near-misses seriously enough. In 1972, after the government had dismissed a barrage of criticism from scientists and others who questioned its dedication to safety, a reporter for the prestigious journal *Science* was moved to remark that "the AEC, in its eagerness to develop a thriving nuclear industry—and to get on with building the breeder reactors it has dreamed about for 20 years—has deliberately bypassed tough safety questions still hanging over ordinary, water-cooled reactors."

Continued criticism forced the AEC to ask a group of Massachusetts Institute of Technology scientists to study nuclear power plant safety. After a two-year investigation, the scientists said that the chances that a nuclear reactor accident would kill 100 people were smaller than the chances that an airplane accident would kill a similar number of passengers. The MIT team added that there was only one chance in a million that a nuclear reactor accident would kill 1000 people.

But as far as I am concerned, statistics like these are quite unconvincing as well as beside the point. I strongly suspect that for every accident the Atomic Energy Commission reports to the public, there are many "near-accidents" and "lesser accidents" that go unreported by these high officials. And I would guess that for every actual or potential accident known to but not reported by high-ranking officials in Washington, there are numerous mishaps that are concealed by lower-echelon workers at the plants out in the countryside. I have worked in industrial laboratories long enough to know that the operators are human, so these "minor coverups" are bound to be frequent. As fission reactors, breeder reactors, and fusion reactors proliferate, there are bound to be more and more accidents, both major and minor. And as these accidents mount in number, the odds in favor of a major nuclear catastrophe will become larger and larger.

Moreover, comparing the two probabilities—that a nuclear

reactor accident will kill 100 people and that an airplane accident will kill the same number—is an example of the kind of slippery reasoning that pervades this whole area of controversy. What everyone knows in his bones, of course, is that an airplane accident, however tragic it may be for the victims and their loved ones, can only kill a few hundred people *at most,* whereas a nuclear accident can, if it is bad enough, kill millions of people, contaminate whole populations, and make large cities and major geographical regions uninhabitable.

Most scientists who have given clear thought to the implications of our drive to develop and commercialize nuclear power agree that almost every aspect of nuclear energy is likely to present us with more problems than solutions. "Alvin Weinberg," Dr. Handler told the Annual Convocation of Markle Scholars, "has depicted the worst possible case by examining the requirements if, a century hence, all primary energy worldwide were to be nuclear. If projected requirements for power were realized, it would be necessary to construct during that period 3,000 nuclear parks each consisting of about eight fast breeder reactors with each park capable of producing 40,000 megawatts of electricity when working at 40 percent efficiency. But that would mean putting four reactors on line each week for the next century and also replacing those that wear out, an absolutely staggering task. When one adds the nightmare of the existence of the 15,000 tons of plutonium required for that many breeder reactors, the health hazards in handling plutonium, the police effort required so that no plutonium is removed for the construction of illicit nuclear weapons, and the task of waste disposal, one need not invoke the possibility of a catastrophic accident to consider that this is an insupportable scenario. Somehow, the world must skip the breeder reactor and go from petroleum and coal to fusion and/or solar energy or it is inconceivable that the human race will avoid a worldwide calamity on so large a scale as to jeopardize the continuing future of our species."

Clearly, the drive toward world-wide industrialization has brought us—every nation on earth—to a time and place that call for some far-reaching decisions. One very important question we must face up to is whether we should go to the trouble of installing nuclear reactors if, in addition to thermal pollution, they will also rain down upon us intensified radioactive pollution problems.

What is the sense of implementing technologies that will endanger our health and create thermal pollution, when solar energy can safely and easily supply us with ten times more nonpolluting energy than all the thermally polluting energy we will ever be allowed to use? Why not go directly to solar energy?

Many energy experts counter that we should in fact move as quickly as possible to develop fusion reactors because they will be both highly safe and highly efficient. They will, it is said, produce far less waste radioactivity than any kind of fission reactor, and at the same time they will generate less waste heat than any type of power-producing technology in existence today. But when a nuclear power expert says a fusion plant will be safe, he means it will be some ten or one hundred times less productive of long-lasting radioactive pollution than a fast-breeder fission reactor of the same power capacity. It will still necessarily be far, far more productive of radioactive danger than any fossil fuel, geothermal, or solar energy system would be. And it will also, therefore, pose correspondingly greater dangers as a target for terrorist activities.

And will fusion reactors produce less thermal pollution

than other nuclear or fossil fuel systems? True, fusion reactors might be more efficient. They might produce less "waste" heat. But this line of reasoning masks a great fallacy: that waste heat alone is responsible for thermal pollution. The fact is that practically *all* energy taken from sources found in the earth—oil, coal, geothermal heat, uranium, deuterium—will eventually be thermally polluting. The waste heat—the heat poured out of the smokestack of a factory or the cooling tower of a fission reactor plant—may get into the atmosphere first. But a little later *all* the heat that was used to turn the wheel, manufacture the car, smelt the copper, air-condition the skyscraper, or heat the home will also find its way into the atmosphere and will add to its thermal load. Useful energy as much as wasted energy ultimately warms the air, ultimately makes its contribution to the hothouse crisis.

There is therefore no reason to hope that fusion reactors will save us from the crisis and prevent the inundation of our cities. All the energy they produce for civilization, no matter how efficiently they produce that energy, will ultimately be thermally polluting. In fact, it can validly be argued that the advent of cheap energy (if indeed it *is* cheap) from nuclear reactors—fission or fusion—will speed up catastrophe because the cheapness of the energy will accelerate our power consumption beyond what it would be otherwise. By the third quarter of the next century, just a hundred years from now, there could be hundreds or even thousands of nuclear reactors poised on the landscape and floating on the oceans.

Experts have long proclaimed the economic advantages of cheap nuclear energy. Dr. Alvin Weinberg, former director of the federal government's Energy Research and Development Office, and a dedicated, thoughtful scientist spokesman for the proponents of nuclear power, has described the "Faustian bargain" made by "nuclear people" with society. He says: "We nuclear people . . . offer energy that is cheaper than energy from fossil fuel," but then he points out that "the price . . . we demand of society for this magical energy source is both a vigilance and a longevity of our social institutions that we are quite unaccustomed to." This reasoning implies, I believe, that the price of nuclear energy will be high—very high—even if much of that price cannot be measured in economic terms. This price might well include, for example, repeated pressures all over the world to establish police states in the search for social institutions possessing adequate "vigilance and longevity." Even the straight dollar costs of nuclear accidents and antiterrorism controls have not yet been realistically included in the estimates of the economic price that will have to be paid for energy derived from nuclear reactors.

Assessing all these factors, then, I see but one sensible conclusion. We must put our resourcefulness to work developing an alternative source of power, a source that will not poison our air or our water, a source that will not heighten our exposure to the threat of radioactive terror, a source that will not thrust us across the threshold of thermal pollution: the sun itself.

7
POINT/
COUNTERPOINT

The assertion that man's accelerating consumption of energy could release amounts of heat into the atmosphere sufficient to melt the ice caps and flood the coastal areas of the world tends to send most climatologists (and other experts) hastening to those blackboards that seem so essential a part of every scientist's office. Chalk moving furiously over gritty surfaces, many scientists will dash off equations, graphs, and charts demonstrating just how shaky most predictions of climatic crisis are.

Calculations that presume to forecast how the atmosphere

will react to heat generated by man's activities, the skeptics say, are based on scientific work that depends on far too many assumptions and too few facts to be taken seriously as the basis for far-reaching forecasts and predictions. Scientists who deal in most biological, chemical, or physical problems can conduct straightforward, conclusive laboratory experiments to prove or test their hypotheses. Many of the problems and mysteries that mark these sciences lend themselves to analyses that can be conducted entirely with equipment ranging from simple test tubes to complex devices that detect and analyze the minutest of reactions or changes in the materials mixed in those test tubes.

On the other hand, as the skeptics point out, researchers why try to decipher all the nuances and enigmas of atmospheric actions do not have the same luxury. Climatologists can create conditions in their laboratory apparatus that are crude approximations to the conditions sometimes observed in the atmosphere outside the building where they work, and they can fly around the world to study weather patterns of every kind. They can send up balloons to gather information in the higher strata of the atmosphere. They can even launch satellites with men and instruments trained and designed to study the atmosphere from outer space. But climatologists, in their attempts to understand just what goes on in the atmosphere, have to depend heavily on "models" to round out their insights. These are not wood or plastic or even gaseous stand-ins for the blankets of air and ocean that surround the earth, but rather mathematical representations of the atmosphere, lands, and seas.

If a climatologist wants to know what effects a change in a local source of heat, say, might have on the weather of a particular region, he will "construct" a mathematical model that includes as many expressions representing known fac-

97

tors—atmospheric temperature, atmospheric density, cloud cover, all at a particular time—as possible. The model will also include equations representing the theoretical relationships whose effects the climatologist wishes to study. The new source of heat, for example, will be translated into a mathematical formulation and written into the model's equations along with the other factors of major significance. Once all the known facts and assumptions are down in suitable mathematical language, the climatologist (or rather, his computer, for almost no climatologist can function effectively without a computer these days) will work through the equations and follow the mathematical logic until he has worked out a solution that tells him what the weather might become under the theoretical conditions he has established.

Climatologists say that the best atmospheric models, the ones that will lead to the soundest conclusions about what changes might come about under different circumstances, are those that have the most facts and the least hypothesized variables written into them. A model with ten facts and one assumed variable will yield an explanation of a presumed atmospheric effect that is far more reliable than a model with three facts and seven assumed variables.

The skeptics say that the models used to predict the disastrous effects of thermal pollution on the world's atmosphere, ice caps, and oceans, are shaky because they include too many assumptions and too few facts. When the eminent scientist M. I. Budyko predicts that thermal pollution could soon be melting the ice caps, the skeptics say he does so on the basis of a suspect model. "There are too many assumptions and not enough givens in Budyko's model," one says. "He has a highly parameterized model that depends on an assumed cloud cover, that assumes how much heat storms will carry, and so on. Since you can't predict these, you have to approximate them, and that is very dubious because the approximations may not be valid at all. Budyko's model has a very limited usefulness."

It is quite true that Budyko's model and my model are too simplistic to permit highly detailed predictions about what will or will not happen to the ice caps. But my point is that the fine details of the prediction are largely irrelevant. What is relevant is something very crude and very basic: If man continues to accelerate his consumption of the earth's buried energy resources at a rate of 4 to 6 percent per year, then in some eighty to one hundred twenty years man will be putting heat into the atmosphere equivalent to 1 percent of the solar rate; after another forty to sixty years he will be putting in 10 percent; after another forty to sixty years it will be 100 percent. Somewhere in that horrendously accelerated sequence—almost certainly within eighty to one hundred eighty years from now—the ice caps will quickly melt.

What I am saying is that man's constantly accelerating consumption of energy *cannot continue unabated* without disastrous consequences: that mankind must level off his consumption of energy when its value begins to rise above the 1-percent and close in on the 10-percent input level relative to the solar rate; that man must begin this leveling-off process very soon! Now! My challenge to those who are skeptical about my conclusion is: Show how it would be possible for man to maintain unchanged the ice caps and the level of the sea even though man's rate of energy consumption becomes greater than about 10 percent of the solar input rate. How could it be done?

There is, it should be said, an alternate side to this coin. The climatologists who fully understand and agree with what I am saying tell me I should not argue that man must level off his growth when his energy consumption rate

99

becomes equivalent to 10 percent of the sun's input power level, for implicit in such a statement is the tacit assurance that it might be quite acceptable for man to accelerate his rate of energy consumption until it does reach that 10-percent level. "If you say that," my allies tell me, "you may be responsible for encouraging people to ignore the danger until it could be too late." So let me be clear: I am *not* saying that there is *no danger* at rates *below* 10 percent. The danger will probably be significant at the 1-percent level. At values between 5 and 10 percent, man's energy consumption rate will probably be causing severe problems. I suspect, however, that man's technological abilities will possibly be great enough to enable him to cope with some of the earliest problems caused by thermal pollution. At rates below 10 percent, for example, the artificial control of the earth's reflectivity might enable man to preserve the ice caps and keep the waters away from his cities. It is when man's power consumption exceeds 10 percent of the solar rate that he will, in my considered judgment, necessarily be melting the ice caps faster than his technological know-how for coping with the results can conceivably develop.

My critics never say outright that thermal pollution is not, or won't be, a very real problem. One group admits that thermal pollution could cause difficulties for mankind. However, they go on to say, the negative effects won't be nearly as bad as I predict. Within the next century there could very well be twenty billion people spread around the globe, all of them using three times as much energy as the average American uses today. But the resultant widespread thermal pollution wrought by twenty billion people using one billion Btus of energy each will leave the world pretty much unscathed. "No doubt something is going to happen," one colleague admits cheerfully. "There might be some climatic changes. The Arctic ice might melt, but that would not be disastrous. We might have some flooding here and there. The Amazon might dry up because of increased evaporation under the stimulus of higher temperatures, and Siberia might grow warm enough to grow sugar cane there. But that's about all. Certainly we are not going to be using enough energy to heat the atmosphere sufficiently to boil the oceans."

However, I must reply that *if man were to succeed in continuing to accelerate his energy consumption rate at 5 percent per year, say, for another two hundred years, he certainly would begin to boil the oceans.* Such a catastrophe cannot actually happen, of course, because the atmospheric heating and ice-cap melting that would come earlier would create such human pain and disaster as to prevent man from ever boiling the oceans.

Note that when a critic makes a statement like "certainly we are not going to be using enough energy . . . to boil the oceans," he has in fact agreed to my conclusion that man will soon have to terminate his energy consumption acceleration. In effect, he has said in a veiled way exactly what I am saying.

A second set of critics admits that thermal pollution may occur, but then they insist that nothing as damaging as I am suggesting will result. At a recent world conference on climatology, Dr. William W. Kellogg, senior scientist at the National Center for Atmospheric Research in Boulder, Colorado, said that the world's weather was warming up. "The reason for this is as follows," he explained to the assembled scientists. "The climate is determined . . . by an overall balance between the solar energy applied to the atmosphere-ocean-ice-land system and the energy radiated back to space as infrared radiation. Anything we do to change this balance will change the climate. We are doing two things on a massive scale now, and the scale is steadily increasing—adding carbon dioxide to the atmosphere by burning the fossil carbon locked in the earth in the form of coal, petroleum, and natural gas, and also adding heat directly by our insatiable demand for energy. Both of these cause the temperature of the climate system to rise."

If in the years ahead twenty billion people consume the equivalent of 1 billion Btus per year, Dr. Kellogg said, the amount of heat released into the atmosphere will be equivalent to 1 percent of the incoming solar radiation. That, he added, could translate itself into an increase of temperature of almost 20 degrees at the poles.

But these temperature increases, Dr. Kellogg continued, will probably not have the consequence of melting the ice caps: "A common theme among 'prophets of doom' is the spectre of the ice sheets melting and the coastal cities of the world being engulfed. Let us look at this idea more carefully. First, the ice sheets of Greenland and the Antarctic are very massive and their tops are several kilometers above sea level, meaning that precipitation on their tops will continue to be in the form of snow even when sea level temperatures are above freezing. Second, their sheer mass means that any significant changes in volume, if they should occur, will be on a time scale of many centuries at least. Third, and perhaps most significant, is the curious fact . . . that the Antarctic ice sheet has apparently grown larger during the warmer interglacial periods than during the glacials, when most glaciers elsewhere were greatly extended. This inverse effect appears to be true of the Greenland ice sheet as well. . . . In summary, it is simply not at all evident that the great ice sheets will melt as a result of general warming, and they may grow even larger, lowering the sea level slightly."

102

But a close reading of Dr. Kellogg's paper shows that he is assuming man will level off his rate of consumption of energy at about 1 percent of the solar rate. If man does in fact do this, Dr. Kellogg and I have little or no disagreement. Maybe the ice caps will melt a bit, maybe they will increase a bit; in either case, I would agree that man can probably cope with the resulting problems, if any. What I am saying —and I think Dr. Kellogg would agree—is: *Man must level off his consumption rate of thermally polluting energy somewhere in the area of 1 to 10 percent of the solar input rate;* and I am also saying: *If man does not do this, if he continues to accelerate his energy consumption rate beyond 10 percent of the solar input rate, then he will inevitably suffer catastrophe through the melting of the ice caps.*

It may be true, as Dr. Kellogg and others suggest, that for a time the increase in atmospheric temperatures, by encouraging evaporation from the oceans, may actually increase snowfall and ice thickness in the polar zones. But even if such ice thickness does increase for a time, it is quite obvious that that increase will be only temporary and won't make any difference in the long run. Ultimately, the atmosphere has to get hot enough to melt far more snow and ice than increased evaporation can deposit on the ice caps.

Furthermore, it is wishful thinking to believe that thermal pollution will finish melting the ice caps only after thousands of years, long enough to give man time to adapt to the consequences. The conventional wisdom among laymen and scientists, of course, has been that changes in the climate are necessarily long drawn out, and that changes in the ice caps cannot occur in less than thousands of years. But my calculations (see note page 14) show that man's constantly accelerating increase of energy consumption will telescope processes that might otherwise take thousands of years into a time span covering less than two centuries. Furthermore, a number of studies of temperature changes within the atmosphere in recent times have suggested that the world's ice caps can, and in fact do, react to atmospheric temperature changes *within less than fifty years.*

Some scientists believe that climatology, though a rapidly maturing science, is still too young to predict with any certainty just what effects man might have on the weather if he goes on accelerating his consumption of energy. In an editorial for *Science,* Dr. Weinberg recognized that the consumption of energy in the world could rise sufficiently to pose problems for mankind. "Until recently," he said, "consumption of energy was expanding rapidly. At the moment, the increase is at a slower rate. However, there are unsatisfied wants in many lands. When it becomes feasible to produce large amounts of energy, the former rate of increase might be resumed and even exceeded. . . . Man was increasing his production of energy by about 5 percent a year; within 200 years, at this rate, he would be producing as much energy as he receives from the sun. Obviously, long before that time man would have to come to terms with global, climatological limits imposed on his production of energy. Although it is difficult to estimate how soon he shall have to adjust the world's energy policies to take this into account, it might well be as little as 30 to 50 years."

But then, having started so bravely, Dr. Weinberg seems to undercut his own arguments. "Unfortunately," he writes, "the science of climatology is unable to predict the ultimate consequences for the earth's climate of man's production of energy."

Of course I must admit that it is, strictly speaking, impossible for scientists to pinpoint the "ultimate consequences" of any train of events. We truly cannot predict anything as accurately as we would like. But just because we cannot foresee every small detail in the development of a problem, this does not mean that we have to close our eyes to what soundly based theory and good common sense state are its potential effects on human life. We may not be able to say that at 11:35 A.M. on June 28, 2093, the second-story windows of the Empire State Building will sink beneath the waves. But certainly we *can* say, as Weinberg says, that within some two hundred years, if the present acceleration of man's energy consumption rate continued, he would be producing as much heat as he receives from the sun. We can then equally well say that at that point the ice caps would quickly melt, sea levels would surge upward, and unparalleled catastrophe would befall the human race. We can say—it is imperative to say—that we should now take the steps that will positively prevent the possibility of such a catastrophe. I agree with Weinberg's statement that we shall probably have to abandon our present energy policies within thirty to fifty years. In fact, given the rigidity of man's traditional habit patterns, I conclude that we must begin at once to try to "adjust" these dangerous growth policies. We must start now to stop the galloping growth of our world-wide power consumption. And it certainly makes no sense to undercut or soft-pedal this important decision for all mankind by dismissing climatology as a science too under-developed to tell us every minute detail we might want to predict.

The last line of defense taken by those who do not want to deal with the implications of thermal pollution is that the vast amounts of heat we will be sending into the atmosphere will probably be neutralized by *concurrent drops in atmospheric temperatures.* This seemingly contradictory state of affairs will come about, we are told, because while some

105

man-made emissions will serve to add heat to the atmosphere, others will serve to decrease atmospheric temperatures. Some scientists say that increasing numbers of atmospheric particles released by industry, automobiles, and trucks, plus dust kicked up by volcanoes or coming from other sources, will scatter increasing amounts of sunlight back into space, thus preventing conversion of the sunlight into heat. The implication that one is supposed to draw, I gather, is that the resulting coolness will offset the warmth induced by thermal pollution, thus enabling us to go merrily on accelerating our energy use.

Others assert hopefully that the vast amounts of heat we will soon be sending into the atmosphere will be neutralized by the onset of the next naturally occurring ice age. These people point out that over the past million years or so, warm stretches of time have alternated, about every hundred thousand years, with colder periods, the periods commonly known as the ice ages. This school of thought says that the current interglacial era is ending, and that a new era of widespread glaciation is now coming on. Thirty or forty years ago, some researchers point out, Baffin Island, in the Canadian Northwest Territories, had areas that were snow free, whereas today these sites are covered by snow banks. Similarly, the ice pack around Iceland has begun to hinder navigation just as it did during the so-called little ice age of the seventeenth and eighteenth centuries.

Plants and animals in the Northern Hemisphere, some scientists say, are already giving signs that an ice age is well on its way. The armadillo, a creature that thrives on warmth, is retreating south after having made inroads into the northern half of the American Midwest. Snails commonly found in the warm forests of Central Europe have disappeared from that area.

Will these projected falls in temperature really suffice to offset man's accelerating heat production and the resulting thermal pollution?

To put it simply: No, they will not.

If a new ice age is indeed coming upon us, it would have to descend as rapidly and as intensely as our heat emissions ascend, to neutralize those heat blasts into the atmosphere. In other words, it would have to fall—crash is a better word —within eighty to one hundred eighty years, and it would have to amount to the equivalent of a drop in solar radiation by as much as first 1, then 10, then 100 percent, then more. That would, of course, be impossible. Actually, most climatologists believe that the drops in temperature leading into a full-scale ice age—drops that could be as much as 30 to 40 degrees in the higher-latitude regions—would occur only over thousands of years, a tiny fraction of a degree at a time, and would correspond to a drop of solar radiation of only a few percent at most. Obviously a crawling descent of the thermometer spread over so long a time would put hardly a dent into man's avalanching thermal pollution effects.

Admittedly, some climatologists now have a less conservative opinion of the time required for the climate to switch from an interglacial to a glacial type. Some recent evidence indicates that an ice age can descend relatively quickly. Ice cores extracted from the Greenland glacier have given some indications that the change from the last interglacial to glacial condition took no more than a few hundred years. Confirmatory evidence for this conclusion has been derived from bottom cores taken from the southern Gulf of Mexico.

But even if the onset of a new ice age were to be compressed into only a few hundred years, the effect would, I believe, still be no more than to delay the consequences of man's exponentially rising use of energy for one or two decades. Dr. Kellogg,

107

for example, agrees that thermal pollution would be unaffected by a new ice age, that man's heating of the atmosphere will probably be strong enough to ride roughshod over any naturally produced declines in temperature. "While . . . estimates of future global warming due to both adding carbon dioxide and direct heating are crude, and there are many uncertainties in the models used, the conclusions seem . . . unavoidable," Dr. Kellogg said at the climatology conference. "Unless there happen to be natural forces in the opposite direction that will act faster than they seem to have done in the last 1000 years, the warming effects of man's activities will dominate in the decades ahead and will produce a global warming that will reverse the present cooling trend."

If theories about man's effect on the atmosphere are fragile, theories about natural oscillations in climate, it must be said, are even more so. When we postulate theorems about what man's use of energy is doing, we are dealing with a good many things that can be measured. We know how many Btus of energy a year man is now consuming. We know how much heat pouring into the atmosphere this represents. We can calculate roughly how long it will take for man's emissions of heat to reach the poles. We also know just how much heat it takes, first to warm and then to melt a cubic inch of ice, and we can therefore calculate approximately how much heat will be required to melt thousands of cubic miles of ice.

On the other hand, although we can measure what is going on naturally in the atmosphere today, we can only guess at what nature did—or why she did it—a thousand, ten thousand, or a million years ago. We can only guess at the various influences that may have been behind the various warming and cooling periods that have passed across the earth during the past million years. Furthermore, these guesses,

108

even the most educated of these guesses, cover occurrences spread over thousands of years. Our tools for gaining insight into the past are not yet so fine that we can really know what happened or why it happened from one year to the next in ages gone by.

Because our knowledge of nature's past is so incomplete, I believe that our projections about what nature is going to do next are the merest fabrications. Just because there is some evidence that a sequence of temperature fluctuations occurred over the past million years or so, there is no certainty that that sequence must continue. Just because there have been temperature drops in previous ages, we should not look to them to save us now from our lack of commitment to an intelligent energy policy. It is nothing but silly for us to sit here and say, yes, thermal pollution seems destined soon to flood our cities and destroy our civilization, but then, an ice age might save us.

Protestations that thermal pollution will be a minor nuisance, or that there may in fact never be a problem at all, are indefensible in the face of evidence that man's activities are already having a significant impact on the climate of the earth.

No one who has kept an eye on the development of civilization and on planetary weather can deny that over the course of his brief history as a higher species, man has moved with gathering acceleration along the path toward inadvertently interfering with the atmospheric goings-on around him. Early in his development, the influences he exerted on the climate were relatively small. The fires he built and the relatively tiny industrial activities he undertook spewed soot and ash into the atmosphere. There the particulates lingered and reflected and scattered sunlight back into outer space before it had a chance to become heat.

109

While some of these processes had a cooling effect, others had the opposite effect. But man was cutting down forests, plowing up grasslands, building cities. In doing so, he was changing the earth's ability to reflect away some of the sun's radiation before it was converted into heat. Even his efforts to irrigate previously arid areas led to localized increases of temperatures, because the evaporation of imported waters held additional sunlight and consequently additional heat in the irrigated areas, thus increasing slightly the local hothouse effects and so the local temperatures.

As man has developed, his activities on earth have increased. Some of these activities have had a cooling effect, but, on the whole, the result has apparently been a slight warming of the atmosphere. Man's agricultural efforts have come to cover a substantial portion of the land—about 3 percent of the earth's total surface. These have diminished the earth's ability to reflect sunlight back into space, thus warming the earth. Man's industrial activities have also become more intense, his urban areas more pervasive. These centers of activity are now the very areas where a good deal of incipient thermal pollution is already being generated, where climate changes are already being observed.

Daily minimum temperature readings taken in urban areas and in rural areas near them show that the urban sprawls are 6 to 12 degrees warmer than the rural meadows. From the small island of Manhattan, more than six times as much energy at the present time is given off to the atmosphere and waters thereabout than is received from the sun. Moscow, Russia, contributes more than three times as much heat to the atmosphere as it receives from the sun; Sheffield, England, gives off more than 1.4 times as much heat as it gathers in the way of solar energy.

Bland assurances that thermal pollution is nothing to

worry about are particularly irresponsible because, soothed
by these assurances, it is all too likely that neither economists nor energy experts, neither politicians nor the public,
will take the time and trouble during the next one to two
hundred years to think out a plan for leveling off the world's
energy consumption rate. Then, when the inevitability of
catastrophe finally becomes apparent, mankind will probably be forced to depend on luck and last-ditch, patchwork
measures. As it becomes obvious that an ice age will not
develop in time to offset the increasing tide of thermal pollution, as oceanographers document ominously accelerating
increases in the levels of the sea, as climatologists become
increasingly ready to admit to themselves and to the public
that they are detecting slight but disturbing decrements in
the ice cover on and around Greenland and Antarctica, the
scramble to put eleventh-hour brakes on the rate of emissions of heat into the atmosphere will start.

Attempts will be made to produce and use energy more
efficiently, thereby cutting the use of fuels and lessening the
amount of heat emitted year by year. Much of the energy
in fuels shipped to various plants is now (and no doubt will
be for some time to come) wasted, spilled out into the atmosphere without making a contribution to the production
of useful power. In refineries where crude oil is processed,
blast furnaces where iron is smelted from ore, and plants
where fuel is used to create steam and drive the turbines to
produce electricity, anywhere from 10 to 90 percent of the
heat produced by the conversion of fuel escapes without ac-
complishing anything productive.

Engineers will therefore begin to insulate furnaces where
energy is used for the conversion of raw materials into other
products. Scrap metal will be recycled in preference to smelt-
ing new ore. In electric plants, heat will be used not once but

111

several times in a sequence of cascaded steps. Three tiers of turbines, not one, will be installed to produce electricity. The very hot exhaust from a first set of turbines can be used to raise steam to drive a second tier, and the exhaust from that group can vaporize a fluid to drive a third set of turbines.

Where heat cannot be used more than once right at the plant or factory, it may be shuttled to nearby homes, offices, or stores to provide them with the heat they formerly obtained by burning natural gas or some other fuel. If this approach were used on a global basis, the world's total consumption of fossil fuels and nuclear energy could be considerably reduced.

The most glaring examples of our wasteful and disorganized approach to energy are the skyscrapers we are building in every city of the world. New York's twin-towered World Trade Center is designed to use more energy than Schenectady, New York, a city with a population of a hundred thousand. Some of this energy is used to light the two 110-story skyscrapers, but much of it is used to provide those who work in the World Trade Center with comfortable working conditions. As in almost every modern skyscraper today, both the Trade Center heating plants and its air-conditioners are kept working simultaneously the year around. Although the two systems naturally "fight" each other to some degree, so that the process wastes energy, it does succeed in supplying either warmer or colder air on demand, summer or winter. And it does this job no matter what the outside weather may be. If the day is hot, the system uses more energy for air-conditioning; if the day is cold, the system uses more energy for heating. The Trade Center's "internal climate control" system not only "fights" itself, it also "fights" what nature is providing in the form of weather-borne energy.

112

Because we are passing through an energy crisis, there is already some talk about making our skyscrapers and homes more efficient consumers of fuel. Yet even at a time when the conservation of energy is being given a bit more consideration than usual, the discussions do not often extend beyond suggestions that more insulation be used in homes or that skyscrapers be given white façades to reflect the sun's rays, thereby diminishing the need for air-conditioning. There is no hint that commitments should be made to alter radically the way skyscrapers are warmed and cooled. There is little doubt in my mind that as soon as the current concern for energy conservation passes, and as soon as it is obvious that we will have as much energy as we need, we will witness the resumption of inefficient building practices and high energy consumption in skyscrapers and homes.

When the threat of thermal pollution becomes undeniable, belated efforts will again be made to cut down on energy consumption in these buildings. Heating furnaces will be run only in winter and air-conditioners will be run only in summer. "Heat pumps" will be introduced—refrigerator-like devices that will themselves consume relatively small amounts of energy in order to push larger quantities of ambient energy from the colder outdoors into the warmer buildings in the wintertime, and to push energy in the opposite direction in the summertime. Even more important, efforts will be made to adjust the temperatures in buildings by taking account of the movements of the sun. Adjustable arrays of panels or blinds will be installed in many buildings to take advantage of the way the sun's rays approach them. In summer, the panels on the sides of the buildings facing the sun will be slanted at angles to prevent the sun's rays from entering them and warming the buildings. Since excess summertime heat will build up inside the buildings anyway, panels on the sides away from the sun will be left open to

113

allow as much heat as possible to radiate away. In winter, on the other hand, the panels facing the sun will be opened while those away from the sun will be closed, for in this way as much natural heat as possible will be gathered from the sun and this heat will be retained as much as possible. Summer and winter, systems of panels will open and move and close as the sun's rays shift direction across the façades of the buildings.

More efficient use of fuels, the use of insulation, recycling of heat, and cutting back on heating and air-conditioning will reduce somewhat the amount of energy produced and used, and thus will reduce the total amount of heat released into the atmosphere. But unless the acceleration of energy consumption is cut back very drastically (or, unless the use of nonthermally polluting energy is finally accepted, as proposed in the next chapter), the savings will not be very great. At most, by converting fuel more efficiently and by reducing the amount of raw materials we use to produce energy, we might give ourselves an additional ten or twenty years of grace.

As thermal pollution accelerates, and global temperatures and sea levels begin to surge higher, scientists and technologists called in for consultations by desperate governments will suggest additional steps to reduce the problems engendered by the earlier, helter-skelter attitude toward energy. The experts might recommend that attempts be made to cut down on the amount of solar energy falling to earth. They might suggest that this be achieved by projecting hundreds of billions of tons of dust into the upper atmosphere, for example, on the basis that the dust would reflect more of the sun's incoming short-wave-length radiation back into outer space. It might also be recommended that the atmosphere in the polar zones be seeded with nucleating

114

agents to increase snowfall and the growth of the ice caps.

On a more practical level, the oceans, even as they are swelling, might be used temporarily to offset some part of the effects of the heat produced by man. Water 1000 to 3000 feet below the surface of the tropical ocean is as much as 30 to 40 degrees colder than the water on the surface of the seas. This colder water could be forced up to the surface through a multitude of hollow pipes or other systems. The upwelling would probably require very little power—less than one horsepower per thousand tons of ocean water per day—and the energy present in the waves or winds at the surface of the ocean would be much more than sufficient to power the upwelling pumps. Once at the surface, the cold water might cool the atmosphere a few degrees. Even more important, the cold water might create dense and highly stable banks of fog capable of reflecting much of the incident solar radiation back into space. The melting of the ice caps might then be delayed ten or twenty years.

Unfortunately, all these efforts will give relief for no more than a few years. By the time the danger of thermal pollution is obvious to everyone, we will be producing heat equivalent to several percent of the sun's annual input of energy into the atmosphere. Thus the use of airborne dust or fog to protect the ice caps, for example, would first have to reflect several percent of the sun's heat back into space. But since a 4- to 6-percent yearly increase in man's use of energy doubles heat output into the air every twelve to eighteen years, less than twenty years after the first planeloads of dust were flown into the atmosphere and the first banks of fog were raised upon the tropical oceans, we would be pouring heat into the atmosphere at double our previous rate. Therefore more dust would have to be flown up into the air and more fog would have to be generated. Just twelve to eighteen years later, we would be adding twice as much heat again, so more dust and more fog would be called on to reflect back 100 percent, then 200 percent, then 400 percent, of the sun's energy—an obviously impossible requirement.

Many such schemes as these could not get very far anyway, because even in the face of rising oceans they would be neither politically nor economically acceptable. Flying dust into the atmosphere would be very costly, and the dust particles would probably settle slowly back to earth, thus reversing their original function while at the same time aggravating hay fever, asthma, and emphysema. No doubt people living far from the oceans would protest the cost and the increased incidence of respiratory diseases in their areas resulting from the desperate attempts to help people living by the shores.

To say that cutbacks in the use of fuels available to us will not work, and to say that technological attempts to offset man-produced heating of the atmosphere will not do much good, is not to say that there is nothing we can do to keep the oceans out of our cities. There is, in fact, a great deal we can do. We can start by being careful about the projects we undertake—even those that do nothing more than change the reflectivity of the earth in one area.

"Today, only a small minority of scientists is concerned, and so plans involving the transformation of landscapes are made and executed without any clear understanding of the consequences that might follow," Dr. Hermann Flohn has written. Flohn, a German climatologist who has studied the effects of many of man's activities on the weather, goes on to say: "Any kind of interference . . . can have both favourable and unfavourable effects on the environment. For this reason the greatest caution must be exercised in the planning and execution of projects that might influence weather or climate . . . The situation demands that we exercise the

116

utmost care to ensure that we utilise the treasures of nature and natural energy in the most efficient manner possible."

We can also start work—today—on developing and implementing a world-wide energy-use control program that will enable us to level off our rate of use of conventional fuels at a point below which thermal pollution becomes irreversible. We can start developing—in earnest—methods that will allow us to predict and to some extent control the global climate, to exert some influence on global atmospheric temperatures. Even more important, we can start now to develop in earnest an alternative source of energy, solar energy.

The ocean that surrounds us on every side, the same ocean that, left to itself, might destroy our civilization within the lifetimes of our great-grandchildren, can be a fruitful converter of that solar energy into the food and the nonthermally polluting fuels that man so greatly needs.

8

SOLAR ENERGY AND OPEN-OCEAN FARMS

In the summer of 1974, just off the island of San Clemente—some sixty miles out in the Pacific Ocean west of San Diego, California—an old eighteen-foot whaler bobs gently on the ocean's sunlit, shimmering surface. Near the ancient vessel, four divers swim silently some 40 feet below the surface of the ocean. Working methodically with slender, foot-long knives, the divers make their way through a dense, shadowy undersea forest of giant kelp, first separating the 100-foot-long plants from the rocks to which they are clinging, and then "sewing" them (by use of huge needles threaded with heavy-gauge nylon fish lines) to a long rope anchored nearby.

After fifteen of the huge plants have been sewn to the line, three of the divers break to the surface as the fourth, a young marine biologist, ties the rope to the back of the boat. When she has finished her task, she climbs aboard with the others and the boat, slowly towing the rope with its attached plants, makes its way to a curious site about a mile out: a bed of kelp growing in the open ocean on a submerged, horizontal mesh of plastic lines supported by buoys and maintained by anchors and lines at a depth about 40 feet under the surface. As the boat wheezes to a halt, the divers slip back into the water and untie the rope bearing the seaweeds floating slack in the water. Tugging their limp, undulating burden behind them, they swim slowly to the mesh below, seek out an open spot among the other lines of the mesh, and hook the new line and its huge charges to their new home.

Today that original farm is gone, destroyed by a degradation of the mooring system which allowed a corner of the mesh of the farm to approach the surface, followed by the unauthorized transit of a large ship (registry unknown) past the warning buoys and through the body of the mesh.

But before the little farm was destroyed it had succeeded in indicating several important truths—that the giant kelp plants can live and grow on a mesh at an open ocean site; that the plants can reproduce themselves and generate young, new plants on the lines and buoys of this structure; that the plants can survive and grow despite the tendencies of fish and other organisms to graze upon or attack them.

So that first ocean farm is no more. Indeed, it was never more than a small patch of seaweeds growing on an artificial "bottom" in the open ocean; a unique experimental garden in

119

the sea; a brief but pregnant learning experience destined, I believe, to occupy an honored spot in the enduring lore of the oceans. The project of which it was the vital initial step was begun by the U.S. Navy and later supported by the American Gas Association and the National Science Foundation. The project itself continues. As improved farms based on the knowledge gained with that first open-sea garden become longer-lived and more productive, there promises to emerge a hugely significant twenty-first century agricultural institution: a system of vast farms on the open oceans, farms that in the future might conceivably cover as much as 50 percent of the world's total surface area, farms that might supply mankind each year with billions of tons of foods; hundreds of billions of additional tons of nonthermally polluting fuels (such as synthetic natural gas, synthetic fuel oil, and synthetic gasoline); and tens of billions of tons of fertilizers, plastics, fibers, waxes, lubricants, and all the other products we have come to expect from the petroleum industry. The ocean farms of the future may well accomplish all this—may serve as cornucopias of fuels, foods, and other products—by taking advantage of one of the most abundant and long-lived sources of energy available to us: the sun.

Whether or not solar energy will ever generate this fountainhead of foods and other supplies for man is still a subject hotly debated by many technologists. Most solar energy visionaries tend to focus on the deserts—the southwestern United States, the Sahara, the Arabian Desert from the Red Sea to the Persian Gulf, the Gobi Desert in northern China, and the desert of Central Australia—thinking to cover them with millions of "solar cells" that would capture the sun's radiation and convert it into electricity. Other specialists visualize projecting gigantic solar energy satellites into orbit. The satellites would not only capture solar energy that would otherwise not reach the atmosphere, but would convert that energy into microwave radiation and then beam it to strategically located receiving stations on earth for distribution to power plants, factories, homes, and office buildings.

Solar energy is a subject that is often—in fact, usually—distorted in thought and debate. Many people, even experts, labor under the impression that solar energy can be used only by way of incredibly complex and expensive technologies. Others say in all sincerity that solar energy would possibly be valuable if only there were some way to store it for use after the sun goes down at night. In 1972 the federal government published *United States Energy—A Summary Review,* a document which allocates exactly seven paragraphs out of forty-eight pages to the subject of solar energy —just enough to dismiss it from serious consideration. In the first paragraph, the study acknowledges that energy from the sun falling on the earth's surface is enormous: "Sunlight falling on an area the size of Lake Erie in one day is equal to the energy from all fuels man has so far burned on earth. Such energy is the prime element in plant growth and in animal comfort."

With that admission out of the way, the government paper goes on to wash its hands of solar energy: "The advocates of solar heating and cooling generally concede that solar heat collecting and storage devices are not economic," it laments. "Space craft electricity using a silicon solar cell can generate one watt of power at a cost of $200 or 2000 times greater than conventional generating equipment." It then concludes: "Owing to the highly unfavorable economics of solar cell produced electricity, it is very unlikely that solar energy will become a significant power source for many years." The publication does not define just what it means by "many years." But in an obvious bit of fallacious reasoning,

121

it banishes *all forms* of solar energy from our up-and-coming energy armamentarium just because *one* method of converting solar energy—the silicon solar cell—is still in its expensive, developmental stage.

The solar cell, it is true, does have disadvantages today. When sunlight falls on a silicon solar cell (or something similar, like a cadmium sulfide cell or a germanium cell) the sun's photons create a voltage at the cell's terminals. This voltage, like that which is generated by a conventional dry cell, can drive an electric current through a light bulb, an electric motor, or almost anything else that depends on electricity. But solar cells are still expensive. Even a small solar cell system—for example, one used on a space craft to provide it with electric power—must often contain thousands of small cells at a cost of thousands or even millions of dollars. To create the amount of electric power that the United States generated with conventional power plants in 1970— about 350,000 megawatts—enough silicon solar cells would have to be manufactured (at prohibitively high cost) to cover almost 10,000 square miles, an area as large as one-tenth of the state of Arizona.

A cell in a flashlight or car battery draws energy from the chemical reactions produced by the substances contained in it. When the reacting chemicals are all transformed— "depleted"—the cell has become discharged, has run out of power. Chemical cells, as long as they possess unreacted chemicals, can furnish power on demand. A solar cell, on the other hand, does not use reacting chemicals to produce power. It can function whenever the sun is shining on it. But only then. It cannot provide power when there is no sunshine (or similar light) incident on it.

Because the solar cells cannot hold energy in store for nights or cloudy days, these cells today would be practical only if storage batteries were used with them. Batteries are not, however, the simple solution they might seem to be. They would have to be able to store billions of Btus of electric energy—enough to last a city for several hours or even days of cloudiness. They would therefore have to be very large, very heavy, and very expensive.

Since storage of electricity during sunshineless days poses obstacles to the use of solar cells, and since there are no clouds and almost no day-night dichotomy in outer space, it has been suggested that such cells be placed on special space satellites. This would probably not be practical. In the first place, it is inherently expensive to send anything out into orbit, much less a bevy of giant satellites all carrying enough solar cells to meet the massive demands for energy back on earth. More important, sun-gathering satellites would convert solar energy into a source of thermal pollution on earth because these satellites would be reaching out and gathering and relaying to earth much solar energy beyond that which now determines the average temperature of the globe and the sizes of its ice caps. The energy gathered by the solar cell satellites would actually add to the thermal pollution already being heaped into the atmosphere by the use of fossil fuels and nuclear energy.

One might wonder at the importance attached to high costs in connection with anything so crucial as a "clean" source of energy. "What if it is expensive," some say, "when the lack of pollution and other factors are all so favorable?" But energy lies at the very foundations of every society on earth. All values in a society can be shown to rest ultimately on a cornerstone of adequate energy. All values are swept away if the energy supply fails. If nations are locked in competitive struggles with one another—and no society as yet has been able to avoid such struggles in this striving, evolving world—the nation with access to the cheapest and most abundant sources of energy will generally triumph, sooner or later.

Thus it is that the competitive struggles between individuals and between nations force us to seek the cheapest sources of energy and to consume them at prodigious and ever accelerating rates.

But it would be a great mistake to suppose that solar energy is impossibly expensive in economic terms. People who make this argument seem blissfully unaware that most conventional hydropower installations are energized by solar energy and are generally able to produce cheaper electricity than their fossil fuel or nuclear powered competitors. These people apparently fail to understand that every farm—the whole agricultural industry—is powered basically by solar energy, and that this industry sells its output—fuels for the human body—at a substantial economic profit.

In economic terms, food energy is 100- to 1000-fold more valuable (expensive) than coal or oil or natural gas or nuclear energy. A lump of coal, for example, has more energy in it than an equal weight of bread or meat, but a person can't eat coal because its energy has the "wrong form" for the human stomach to digest. If one could eat coal effectively, one could eat for a whole year at a price less than $5.00— that's how cheap coal energy is compared to food energy. And that is part of the reason why the solar energy conversion systems called farms are able to operate profitably in economic terms.

Solar cells and the ancillary technology they require are expensive and impractical for generating commercially important power at this moment, but it would be foolish to disparage them and imply that they will never be cheap or practical. On the other hand, it would be still more foolish to suggest that the 100 million billion watts of power that come to us from the sun every day must be wasted because solar cells are too expensive.

One difficulty that such schemes as solar cells and solar satellites generate for the whole field of solar energy is what I call the "high-technology" fallacy. Technological experts seem often to fall into this fallacy. Such schemes, and the statements of some technologists, make people think that only the most advanced and expensive of technologies hold any hope of making solar energy available for man's use. The facts are quite otherwise. Man's whole existence was powered exclusively by solar energy until recent centuries —and this by way of the most primitive technologies, of course. Probably more than 90 percent of his power was produced by the sun as recently as 1850 or 1900. In the next few decades he will be forced—for his own good, I do believe—to return to sunlight to satisfy the great bulk of his energy needs.

The solar cell is by no means the only method we have to convert solar power to man's uses. Technologists around the world are working on a variety of ways to make the enormous energy delivered to earth by the sun available to meet all our various needs at relatively low prices. Because the sun does not heat the earth evenly, there are substantial temperature differences within the atmosphere, and these differences in temperature lead to the movement of air—the emergence of winds—from one part of the atmosphere to another. Generations of farmers have taken advantage of these air movements by using windmills to accomplish some of the work they have to do around their farms. Some technologists believe that huge banks of windmills—albeit more sophisticated and larger windmills than those used by individual farmers—should be used to generate power to meet much of our needs for electricity.

Just as solar energy creates temperature differences in the

125

air which lead to the winds blowing from one temperature zone to another, these winds in turn (along with temperature differences in the oceans, temperature differences generated and maintained by the flow of solar energy) drive the flow of gigantic ocean currents. Some energy experts believe that huge turbines, to be turned by the ocean currents, should be built and placed in the oceans so that useful energy for man can be extracted from such currents (the Gulf Stream, in the Atlantic, for example). Other experts suggest that the temperature differences (maintained by the sun) between the warm surface waters and the cold deep waters of the tropical oceans can be used in practical ways to generate electricity and other forms of energy.

In preparation for turning to the sun for our supplies of nonthermally polluting energy, we must recognize that solar energy is not as hard to capture or to store as some want to believe. Strictly speaking, the trees growing in the forest represent stored energy, and so do the lakes behind hydropower dams in the hills and mountains, because the sun is responsible for the evaporation and the condensation processes (rain, snow) that raise and deliver the water to the lakes. The winds blowing over mountains and deserts and cities and seas, the waves generated by these winds on the surface of the sea, the heated surface waters of the tropical oceans, the currents flowing through the oceans—all these represent solar energy that is available for man's use both day and night, year in and year out.

However, I deeply believe that the *best* approach to capturing and storing solar energy for the greatest long-term good of mankind will be by way of open-ocean farming. My reasons are simple and basic. Because the oceans cover 71 percent of the earth's surface, most of the solar radiation that falls to the surface falls on the seas. Therefore, if man is to

126

use this energy, he must go to the oceans to gather it. Ocean farming can capture this energy, and it can then use this energy to generate, directly and indirectly, huge new quantities of food, the most precious large-scale form of energy known to man. And it can do all this at what promises to be very feasible prices. Further, additional parts of the ocean farm harvests can readily be converted into synthetic natural gas, alcohol, gasoline, fertilizers, plastics—all the products customarily obtained today from the petroleum-petrochemical industries. This too can be accomplished relatively economically. Finally, ocean farming possesses great potential advantages over its only close competitor, land farming, in the race to power man's long-term future life. The oceans are far larger than the lands, seaweeds can be more efficient converters of solar energy than land plants, the oceans need no irrigation, ocean plants are never subject to killing frosts, and nearly every location on the open ocean is only a thousand feet or so above nutrient-rich sea water that can readily and cheaply be upwelled for the sustenance of the crops.

Although short-wave photons, those high-powered little packets of sunlight, usually release heat upon impact, they can also bring about other effects. When photons strike a plant, for example, not all the blows of energy they deliver are necessarily converted into heat. Some of them power or activate the processes of photosynthesis in the plant's leaves. As these photons penetrate into the leaves, they are used (in effect) to break up water molecules that have previously been absorbed by the plant from the surrounding environment. Then the plant takes hydrogen atoms split away from these water molecules and somehow* uses additional solar photons to weld those hydrogens to carbon dioxide

---

* Science does not yet know exactly how all the processes of photosynthesis work.

molecules that the plant has also absorbed from its surroundings. Thus the system produces carbohydrates and other chemicals that store the energy from the solar photons in the molecules which make up its body. During this process oxygen molecules are given off into the environment. The plant uses the solar energy stored in the photosynthetic substances it generates to sustain its life and to develop—to extend its leaves, grow its roots, create its flowers and other organs. The energy that is released by the solar photons during the photosynthetic process is bound up—held captive —in the various parts of the plant.

We can take advantage of the photosynthetic processes by assiduously growing vegetation to capture the incoming solar energy. And no matter how, or how much of, that vegetation-bound energy is used by man, it will not—it *cannot*—increase the earth's temperature in a thermally polluting way.

Why can't this energy derived from ocean farms add to the thermal burden of the planet earth? Because today all this cascading sunlight warms the oceans and then (through radiation, convection, conduction, and evaporation) warms the atmosphere. If we increase the total vegetation growing in the oceans until it is able to store 1 percent, say, of the total solar energy that showers down upon the surface of the earth, we will be taking solar energy that would otherwise have served only to increase the water and air temperatures by some two or three degrees, and storing that energy in the vegetation of our ocean farms. As a result, the average air temperature around the earth will tend to drop by that same two or three degrees.

The global temperature will only tend to drop, however, because men will continually be converting the plants into food and fuels. These products will continually be flowing

128

to the world's cities for sale and use. There and everywhere, as the products are consumed, they will continually be releasing their stores of energy and creating a local tendency for the temperature to rise. That increased heat will diffuse through the atmosphere. But it will not raise the global average temperature because it will be neutralized by the decrease in heat previously created over the seas. Thus the total average temperature around the globe will stay exactly the same as if the ocean farms did not exist. The positive thermal effects over the cities will be created hard on the heels of the negative thermal effects produced over the oceans, and these effects will exactly cancel one another. The localized increases and decreases of temperature will have little or no chance to make any kind of impact on the world's overall atmospheric temperature and heat balance.

The ocean farms will also maintain unchanged the global average oxygen and carbon dioxide concentrations. At sea, where the plants grow, carbon dioxide will be consumed and oxygen will be given off; on land, where the foods and fuels coming from the farm system are consumed by man, carbon dioxide will be produced and oxygen will be consumed; on a global average basis, reckoned over a number of months, these effects will all exactly cancel one another and the environmental balance will remain undisturbed, no matter how large these farms become.

One of the desired objectives, of course, is that the ocean's bountiful yield of plants will help us feed the billions of human beings who will inhabit the earth in years to come. Some experts doubt, however, that even the sea will be of much help. They point out that the ocean's fish populations are dwindling quickly. Since many nations, particularly Japan and the Soviet Union, are using mechanized equipment that has made fishing highly efficient, some food scientists warn

129

that the oceans will soon be seriously short of edible fish.

If we continue to insist on just crisscrossing the seas with fishing nets in tow, we will certainly further deplete their supplies of edible fish. But if we make a concerted and well-planned effort to put the ocean under organized cultivation, we will in all probability witness a vast increase, perhaps tenfold or greater, in the world's overall fish population. The fish that until now have been staples of human diets have thrived in areas close to the shores, in those relatively tiny regions of the ocean where natural upwellings have supported an abundant plant life. Where there are plants, the necessary foundation for bountiful life cycles exists. Plants small and large provide food for small fish, and then the small fish attract the larger fish that like to make meals out of their colleagues. For years, fishermen have taken advantage of the near-shore ecological systems to draw their catch out of the oceans easily.

The ocean farms, of course, are not going to increase fish populations by protecting them from the fishing boats. Instead, ocean farms will bring abundant plant life to major portions of the oceans where little or no vegetation grows today, so the farms will provide the basis for the development and growth of thousands of new ecological communities, including fish and other sea animals eaten by man. "One of the important features of the giant kelp is that it provides a surface for animals to roost on," says Dr. Wheeler J. North, professor of environmental sciences at Caltech and the Ocean Farm Project's Principal Scientist. "Scallops and other creatures can encrust the blades of these plants to such an extent that sometimes you can't even see the kelp underneath," he adds. "The ocean is very crowded with respect to available surface where animals can attach, and a kelp bed is like Manhattan Island, where you crowd a lot of people into a small space by building skyscrapers. There are more than 10 square feet of roosting surface provided by kelp plants for every square foot of ocean bottom under a dense kelp bed. Certain animals love the shade provided by kelp. Various anemones, crustaceans, molluscs, and echinoderms (starfish, for example) are shade-loving animals, as are some mid-water animals, and these creatures tend to gather in the kelp. The fishes love kelp too, and they come in and feed on all the little animals hiding in the cracks and crevices, and roosting on the surfaces."

It is clear that the new vegetation will sustain a substantial upsurge of animal life in and around the farms. The fish will still be easy to catch, and there will be a lot more of them.

Several methods are available for harvesting the fish hiding in the kelp beds. For example, gill nets can be used, or the fish can be attracted into nets by flashing lights or electric fields.

Not only will the ocean farms greatly increase the harvest of fish life from the seas, they will also, in all probability, be accompanied by large-scale "mariculture" operations in which the kelp is used to feed desirable kinds of fish, shellfish, and other organisms being raised by the billions in specially built tanks filled with sea water.

Moreover, the ocean farms will produce not only fish but also cattle, sheep, chickens, and other animals suitable for the dining-room table. It is by no means unlikely that our grandchildren and great-grandchildren will be eating beefsteak, pork chops, eggs, butter, milk, drumsticks, and many other foods indirectly derived, partly or entirely, from the seaweeds used as feeds for livestock.

Using the ocean's vegetation directly as a part of man's diet is also a practical idea. The eating of seaweeds goes back

131

to prehistoric times among many of the world's coastal peoples. English colonists who came to start a life in the New World brought with them the recipe for blancmange, a pudding prepared by boiling a red seaweed called *Chondrus crispus* (also known as Irish moss) with milk. Health food stores in the USA and Europe feature numerous kinds of tablets, pills, and food supplements composed partly or wholly of various seaweeds. In Japan today, seaweeds are an integral part of the diet. In fact, the Japanese eat 100 million pounds of various seaweeds each year in their delicious sauces and salads. In the United States and other Western nations, seaweeds are now used in sherbet, cheese, yogurt, chocolate milk, imitation coffee creams, custards, salad dressings, jellied candies, pie fillings, bread, batter mixes, pickled meats, meat pastes, frozen fish, sausage casings, preserves, frozen fruit, beer, candy, and ice cream.

As the world's population increases and as the ability of the terrestrial farmer to support it diminishes ever further, marine plants are almost certainly going to become a direct part of our diets. They might be processed to make up a substantial part of frozen food dinners, canned foods, and cake and cooky mixes. Processed and supplemented by various flavorings, the protein-rich seaweeds could be fashioned into "artificial" hamburgers and steaks. We shall probably be able to process seaweeds into special foods prepared to meet the needs of ill people. Many people suffer from hypertension, heart disease, and iron deficiencies, and marine plants might be used to make better diets for them. Seaweeds could also be helpful to people who have to cut down drastically on their animal fat intake, who must restrict their caloric consumption, or who have to take special mineral supplements every day.

Although work on the ocean farm concept is still in its

early stages, we have already figured out ways by which these farms can be organized, harvested, and even protected from storms.

Most marine plants use their "leaves" (blades) to absorb photons and nutrients from the environment. They use their "roots" (holdfasts) only to anchor themselves to the ocean's bottom and to gain a secure station in life. Therefore, in areas of the ocean where the floor is very deep, we will build artificial bottoms for our crops. These artificial bottoms will consist of far-flung networks of nylon-like lines which will be submerged 40 to 80 feet and to which the plants will be attached. Periodically, ships outfitted with undersea clippers will skim off the surface canopies of the farms, thus harvesting the crops.

Farms close to shore will presumably ship their harvested plants to processing factories on land, but for farms located far out at sea it will probably be cheaper to process their vegetation on giant platforms floating nearby. There, some of the crop will be fed to cattle, sheep, and marine organisms, while the rest of the vegetation will be taken to chemical processing plants also housed on the platforms. In these factories the kelp will be passed through warm water to wash the slippery sulfuric coating off the fronds. Then the crop will probably be chopped up into small pieces, pressed, and passed through hydrochloric acid to break down the cells of the material and leach out the remaining water and salts. After a few more chemical treatment steps, the kelp will be ready to be converted into products useful to man.

Even today, in addition to the role it plays in the production and processing of various foods, kelp is being used to manufacture waxes; lubricants; dyes; solvents; plastics; industrial gums; pharmaceutical products; light oils; industrial chemicals; emulsions for products like face creams and

133

toothpaste; as a component in packaging, fibers, and paper; for cardboard; and for ink and paint. In time, as the ocean farm industry expands and the kelp harvest grows, the seaweeds will make greater and greater contributions to the production of hundreds of thousands of items necessary to our way of life—especially as the raw materials available today (coal, petroleum, natural gas) become depleted.

Some of the soluble sugars pressed out after the kelp has been chemically processed will be fermented to make ethanol, the type of alcohol used in many beverages. Ethanol is also a good motor car fuel. Other plant components will be exposed to bacteria to make methane gas, the major component of natural gas. The bacteria actually digest the plant compounds for their own use, but in the process they rapidly and efficiently produce the gas as a by-product (bacteria do exactly the same thing when they act on food in the digestive tract; we politely call the resulting gas flatulence). Since the bacteria will be doing their work in giant vats, pumps will be used to draw off the methane gas and direct it into storage tanks, pipe lines, and liquefaction units designed to prepare it for transport by tanker ships to distant cities. On our ocean farms of the future, even the sludge that is left over after bacteria have digested the kelp and produced methane gas will find good use: the sludge is high in nitrogen, so it will be valuable as a fertilizer not only on the ocean farms (if we need it there) but on terrestrial farms as well.

Although the engineering problems are truly formidable for any open-ocean structure that is required to endure for a number of years or decades, I believe they can be solved. Because the network of lines will be submerged below the surface of the ocean, for example, storms at sea won't represent a prohibitive danger to our crops. It we find that a very severe storm is in the offing, we could probably submerge the mesh deeper than usual, to keep the plants well out of the way of the roiling waters until the storm has passed by. The plants will also be protected because the farms themselves will probably be mobile. Each farm will be designed to drift a bit with the surface waters, because if the farm were to be made completely stationary, currents would exert a tremendous drag on the mesh, and they might destroy the plants. (We could make the lines heavy enough to withstand currents, but that would probably be overexpensive). Thus if an especially severe storm came along, one capable of creating a relatively rapid surface current, the farm would move along with the surface water. The plants would not be forced to hold their place and run the consequent risk of being torn apart; instead, they would be riding with the current. Once calm is restored and the storm has died down, a relatively low-powered engine, probably deriving its energy mainly from the waves or winds, should be able to move the farm slowly back to its original position.

Before ocean farms can become a large-scale industry, many questions will have to be answered—including what plants will give us the best and most abundant supplies of food and fuel.

The original experimental farm off San Clemente Island was stocked with just one kind of seaweed, *Macrocystis pyrifera*, more commonly known as giant California kelp. We chose to cultivate this seaweed as our first experimental crop for several reasons. This kelp is known to be one of the world's most efficient plants for converting and storing solar energy. It is so efficient, in fact, that it can grow at rates up to an astounding foot or even two a day. And it can reach a length of 200 feet or more. The plant is sturdy and has a natural longevity that is cut short only by the accidents of disease, storms, or an unusually hungry school of

135

sea urchins. The plant's fronds are spread out just below the surface of the ocean, and they are kept floating there with the help of small, air-filled bulbs—pneumatocysts—that grow at the base of each blade. The fronds, plus all other parts of the plant, absorb the necessary nutrients from the water and carry out photosynthesis for the plant. Unlike the other parts of the kelp plant, each frond lives only a short time—six to nine months. But like grass, new fronds are coming along continuously. California's flourishing kelp industry has worked out methods to harvest these fronds as they mature, and then to process them for the already established world-wide market for kelp products. Much of our homework, therefore, has already been done for us. We may only have to adapt the industry's harvesting and processing methods to suit our needs.

Before our experimental farm was begun, Professor North had learned how to cultivate this kelp artificially, a very important step if we are to be able to plant millions of acres of ocean with our crops. Mature kelp plants produce billions of microscopic spores in order to reproduce. Under nature's harsh conditions, only three or four out of every 1000 billion spores survive to grow into full-fledged plants. "Each spore," Dr. North explains, "has two tiny tails that propel it through the water. But it is very slow. It can move only about three feet an hour, and it stays viable for only a day or two. In that period, it has to find a hospitable mooring or die."

Dr. North is now turning much of his attention to making ocean farming a reality. In his role as our chief scientist, he is becoming more and more the apparent prototype of the twenty-first-century open-ocean farmer. A lean, wiry, good-humored man whose face and skin show the traces of an abundant exposure to sun and sea, Dr. North talks and writes warmly about turning the oceans into the new fron-

136

tiers of agriculture. "Almost everything you can say about open-ocean farming is basically favorable," he says. "It's an exciting project, and the further I get into it, the better, the more exciting it looks."

In addition to being enthusiastic about the future of the ocean farm project, Dr. North is also realistic about the work that must still be done if the sea is to support man. "If we are ever going to solve the problem of getting all the people of the world properly fed, we are going to have to turn to the sea sooner or later," he believes. "This is because the area of the oceans—the land that's covered by the sea—is equivalent to fifteen continents.

"It has been said that in order for us to exploit the sea successfully we should stop being hunters and become farmers. But before you can farm a place you have to know what makes the soil rich, why plants grow, what plants are best to grow, and what are the most useful animals which can be handled by man in agricultural endeavors."

Although there is good reason to believe that we will be able to make wide-scale ocean farming a reality, we must, as Dr. North points out, face a number of challenges. We will have to be on the lookout for additional kinds of seaweeds to grow, because *Macrocystis* might not be best for meeting all our needs. Terrestrial farming became increasingly successful over the centuries as farmers learned what crops to use in different soils, when to harvest, how to rotate their crops, how many plants to put on one acre of land. In much the same manner—though perhaps on a more organized, scientific basis than in the past—we will have to learn how our marine crops act under various conditions. Water temperatures vary widely in different parts of the ocean, so we will have to examine how different temperatures affect various kinds of seaweeds. Kelps are native to places like

137

the coasts of Mexico, California, Chile, and New Zealand; will they grow in tropical waters that are substantially warmer, or will they require the coolness as well as the nutrient contents of the upwelled waters in the tropical latitudes? Will we be able to grow our plants—whether kelp or other types of seaweeds—closely packed together so as to extract the maximum amount of vegetation from each acre? Or will the various seaweeds need a lot of elbow room if they are to be able to resist diseases and grow well?

"My guess at this point is that at least two or three kinds of seaweed will be cultivated," Dr. North says. "In tropical, warmer waters, we may want to grow a type of plant known as *Sargassum*. In colder waters we might grow a seaweed known as *Alaria,* which can grow in waters as cold as 30 degrees Fahrenheit and even in many regions where ice forms."

Just as successful terrestrial farmers have learned to increase per-acre yields by using fertilizers, marine farmers who will be sowing the oceans will have to understand and fulfill the nutritional requirements of their plants. Equally important, they will have to learn how to hybridize and otherwise manipulate the basic genetic qualities of their plants and animals so as to increase their yields and their resistance to diseases and parasites.

Conventional wisdom has it that the world's oceans are teeming with life, that there is a plethora of animals and plants in every cubic inch of sea water. In reality, only in the relatively small coastal areas, where tons of organic materials are deposited by the rivers and streams pouring down from the continents, or where natural upwellings occur, have sufficient nutrients been available to support an abundance of shellfish, fish, and plants. Furthermore, only the coastal areas are shallow enough to allow plants like kelp to attach themselves to the bottom and still reach, by growing upward, the surface of the water and the all-important sunlight.

Nutrients from the rivers and streams do not readily find their way to the parts of the oceans that are thousands of miles from shore. To be sure, there are always some thin, highly diluted nutrients floating about in those far-off areas of the sea, and there are always some scattered microorganisms feeding on those nutrients. But when the microorganisms die—plant and animal alike—they sink to the lower depths of the oceans, carrying what nutrients they have absorbed with them. Thus the downward drift of these organisms is continually depleting the nutrients contained in the surface waters of the open oceans. The bodies of the microorganisms dissolve in the depths of the ocean, releasing organic and mineral nutrients there. Therefore, vast quantities of nutrient materials collect hundreds and thousands of feet below the surface of the water. Because sunlight does not penetrate to the depths where the greatest concentrations of nutrients are found, no effective plant growth can take advantage of the stores of sunken nutrients. And since the waters of the oceans turn over only very slowly under natural conditions, it may take thousands or tens of thousands of years for these nutrients to get back to the surface, where growing things can use them.

One of the major technological challenges of the project might be to make the deep-water nutrients available to the large plants we will import and anchor to the artificial bottoms of the ocean farms. This might be achieved, for example, by pumping water that lies a thousand feet deep or more up to the surface through a hollow pipe placed in the middle of each farm. From this central pipe the nutrient-rich water might be pushed into hundreds of horizontal tubes

139

laid out to crisscross the farm at depths of 40 to 80 feet. These tubes would function as structural supports for the plastic lines to which the kelp plants are attached, and they would be pockmarked by small holes in order to release the cool, nutrient-rich waters to the plants.

The mechanisms for bringing up the deep water should present few problems. Even a pump as small as a single horsepower should be able to do the job for each acre of farm. Since 100 horsepower or more is usually available in the form of "swell" (long-period waves) on each acre of open ocean, it seems probable that the upwelling pump for the ocean farm can be a simple float whose up-and-down motion drives a conventional, lift-type pump roughly like those formerly used in farmhouse kitchens. Oceanic winds can also be used to power the upwelling system. Some fuel from the farm can be stored for use in powering the upwelling pump on those rare occasions when both waves and winds are insufficient. (Probably less than 10 percent of the farm's output of fuel would be necessary to operate the upwelling pump continuously).

Since the upwelled water will be colder than the surface water natural for the tropical and temperate zones, the upwelling system of the ocean farms can possibly be useful in helping to offset the overheating of the earth's atmosphere that results from thermal pollution. Though this effect would not be sufficient to permit man to go on accelerating his energy consumption rate for any long period of time, it might be man's quickest, most practical, and least expensive way to offset temporarily an excess of global thermal pollution.

In the absence of any thermal pollution, of course, the upwelled water might represent a climatic hazard because of its low temperatures. To forestall adverse results, we will

140

have to determine carefully how we can get the best mix of nutrients and cold water up to serve our plants. If we find that we cannot bring up adequate amounts of nutrients without also bringing up unduly large quantities of cold water, we may have to find ways to adapt conventional fertilizers to our ocean farm needs.

It is, of course, easy to talk theoretically about converting marine plants into food for cattle, sheep, fish, and even people. Much more research on nutrition and digestibility is needed. It is a good omen, however, that the first scientific studies done on feeding kelp to sheep indicate that these animals—and therefore probably cattle as well—can digest *Macrocystis pyrifera* about as well as they can alfalfa.

Since man is an animal, many such questions will have to be answered before we can serve him foods processed from ocean farm plants. Many seaweeds are rich in the nutrients essential to people. One species of seaweed, called nori (one of the seaweeds used extensively in Japan), makes almost 75 percent of the protein it carries available to human digestion. Other seaweeds offer large amounts of vitamins, including A, B-1, B-2, B-12, C, D, E, niacin, and pantothenic acid. But we don't yet know how much, if any, of the human dietary requirement can usefully be filled directly by the giant kelp plant.

Finally, even if we find adequate ways to convert seaweeds into foods for human consumption, and even if we determine that seaweed products can make up a substantial and beneficial part of the human diet, we may discover that there are many people who do not have the right enzymes in their digestive tracts to metabolize all the nutrients present in the seaweeds. Some nutritionists believe that Asians can make assorted seaweeds a steady part of their diets because adaptive mechanisms worked out by evolution over

141

thousands of years have given them the ability to produce the enzymes necessary for successful digestion of marine plants. A major research effort might have to be directed to devising ways that would enable other people to digest seaweed products if they cannot do so on their own.

To make absolutely certain that no significant problem is overlooked—including the choice of crops, methods of fertilization, and procedures for harvesting the seaweeds and converting them into foods and fuels for human consumption—we are now in the early phases of a three-stage, twelve- to fifteen-year program to work out all the technical and economic factors that bear on the open-ocean farm concept. This project is being managed by the Naval Undersea Center in San Diego, California, for the U.S. Government, and the American Gas Association.

Because the Atlantic and Pacific oceans differ significantly in their current-flow and nutrient characteristics, we want to establish a small farm in each ocean to learn just how much nutrient-rich water we have to add (and how to add it) to our crops under widely different conditions. The Atlantic farm, placed near the powerful Gulf Stream, may teach us how currents will affect, positively or negatively, our seagoing farms. Also, since the Atlantic is subject to heavier traffic than the Pacific we should get some preliminary ideas about how to keep farms and ships apart if ocean farms come to occupy more and more ocean space. The Pacific will also have its lessons to teach us. The Pacific is far bigger than the Atlantic, so our Pacific Ocean farm should tell us how to ship and distribute marine farm products efficiently when the points of production are far from the points of consumption.

If these small preliminary farms answer our questions and

142

encourage us to believe we are on the right track, we plan to move into the second stage of the project by establishing one or two large farms, each of about a thousand acres. These farms will not, we hope, present us with new problems, but they will teach us how to solve some of the same challenges on a grander scale.

Finally, if these 1000-acre farms are successful, we visualize moving into stage three and establishing, by 1985-1990, a 100,000-acre farm to help us work out the final problems of bringing full-scale organized "aquaculture" to the open oceans. The 100,000-acre farm will, in all probability, serve as an adequate springboard for efforts to start farming the open oceans on a commercial basis. Once the 100,000-acre farm is running smoothly, the total cultivated area of the ocean might be increased by some 10 or 12 percent each year. Then, by the year 2100, we could possibly have some 30 to 70 percent of the world's oceans under cultivation.

Neither Dr. North nor I—and none of the other scientists we have consulted for advice on the project—can foresee any insoluble technological problems. But of course that does not mean that the road is wide open to a concerted effort to cultivate the seas.

When we talk about converting the ocean into a vast source of ever flowing energy for the world, raw materials for industry, food for people, we are talking about exploiting a region and a resource that is already a politically volatile issue. Until the mid-1960s the nations of the world took the oceans pretty much for granted. But within the past few years more and more nations have decided that their territorial rights extend two hundred miles out to sea. In part, governments have done this to assure that only their own fishermen shall have access to the dwindling supplies of fish near the shore. But they have also extended their jurisdiction

143

because there is an increasing belief among geologists that the ocean floor is a storehouse of important minerals. Governments therefore want to bring as much underwater territory under their control as possible.

I believe that, because the photon-laden surface layers of the oceans can generate huge amounts of vegetation for food and fuels and other products, they will soon be seen as potentially even more valuable than all the precious ores and resources buried in the ocean bottoms. If they are, this may well intensify the race among the nations to establish sovereignty over defined and militarily defensible regions of the seas.

Within one hundred fifty years or less, systems of ocean farms will probably be called upon to supply 20 to 50 billion people with a major portion of their energy and food supplies. How will these farms be owned and controlled? How will the benefits of the farm systems be split up and parceled out? Every country with a border on the world's oceans may succeed in asserting that its national interests extend two hundred miles out to sea. but who will then command the fruits of the remaining billions of acres of ocean surfaces and bottoms? What will be the situation of nations with little or no access to the oceans?

Many small, island-type nations may have neither the technological nor financial resources to establish ocean farms on the relatively huge, two-hundred-mile-wide zones of ocean they may possess. At the same time, larger nations with limited coast lines may have the necessary funds and technological skills to be able to fill their two-hundred-mile zones with farms, but the farms may be too small to meet the needs of the populations. A highly nationalistic approach to ownership of the oceans might very well mean that substantial parts of the ocean would become fields of battle rather than fruitful farms for man.

144

If we are to deploy a successful network of ocean farms, we may have to decide that the old concepts of territorial jurisdictions by sovereign nations are not the best way to garner the fruits of the sea. Maybe the sea, like outer space and the moon, should be regarded as the possession of collective mankind. To use the oceans in an equitable way for the benefit of every country—including those that have no outlet to the sea—we might need and want to establish an international authority with the right to license the installation of ocean farms. Or the nations might establish an international authority with the power to formulate laws and to police activities that use the ocean. Finally, mankind might decide that the best possible alternative is to form a worldwide government, to manage not only the oceans but also the earth's climate and material resources in the best interests of all peoples.

None of these solutions is without its difficulties and drawbacks. An international authority with no power save to hand out licenses may not be able to prevent corporations and nations with great financial, technological, and military powers from monopolizing vast portions of the seas. On the other hand, an international authority with sufficient power to regulate the whole world's aquaculture activities would probably not be acceptable to powerful countries fearful of subordinating their power (and their advantages) to a flock of smaller nations. And to expect 150 nations comprising 20 to 50 billion people to reconcile their political differences sufficiently to form a united global nation may, though necessary, be totally unrealistic.

It may be the scientist-dreamer in me talking, but I somehow feel reasonably sure that if men are given enough time they will find the technological knowledge, economic means, political arrangements, and—perhaps most important—the moral willpower to farm the oceans for the common good

145

of mankind. Ocean farms, I believe, can yield enough food to feed roughly 3000 to 5000 persons for each square mile of ocean that is cultivated. At the same time, each square mile of ocean farm area can be expected to yield enough energy and other products to support more than 300 persons at today's U.S. per capita consumption levels, or some 1000 to 2000 persons at today's world average per capita consumption levels. The oceans contain (conservatively estimated) some 50 to 70 million square miles of cultivable surface area. If we undertake a systematic development program for the oceans, I believe that by the year 2000, ocean farms could be supporting some 750,000 people; by the year 2050, some 100 million people could be basing their lives on the products derived from the ocean farm harvests; and by the year 2100, we could be farming 50 to 70 million square miles of ocean (35 to 50 percent of the total ocean area) and supporting the earth's projected total population of 40 to 50 billion people with the greater part of their daily food and energy needs. A well-developed system of ocean farms should be able to provide all the food, energy, and material requirements of up to 200 billion people, assuming they live at today's world average consumption levels, for as long as the sun continues to shine at about its present rate.

9
EPILOGUE

I asked a scientist friend whose judgment I value to read this book in advance of publication. "My principal difficulty," he commented, "is putting aside my worries about economic collapses, nuclear wars, and other failures of the social system, in order to focus my concern on melting ice caps after 2055. I remain preoccupied with disaster threatening in my lifetime."

His concerns are well justified and widely shared. Before most of us die, all-out nuclear war may ravage the planet; if not, drought, bad agricultural policies, and soaring populations appear sure to bring on hunger, malnutrition, and starvation on a scale far greater than the worst famines ever witnessed before. Democracies may falter and dictatorships of the right and left, rising on empty promises of simple solutions to complex threats, may spring up all over the globe.

Nevertheless, the only certain and unavoidable catastrophe destined surely to end man's present growth phase, even if he is so clever and wise as to succeed in avoiding all other earlier catastrophes, is that of thermal pollution. This is the brooding giant standing at the end of the growth road man is now striving so mightily to pursue. This is the catastrophe that must be understood if the others are to be seen in their proper perspective. This terminal catastrophe to man's present attitude toward life stands eighty to a hundred eighty years in the future—no more. The only way to avoid that catastrophe is for all mankind to leave the growth road soon.

We must understand too that although the final flooding of the cities may come only in the lifetimes of our great-grandchildren, the triggering of the melting—the turning point after which a possibly irreversible chain of climatological events could lead to the destruction of the polar ice —will almost certainly come within the lifetimes of most people living today. Even so persevering a proponent of nuclear energy as Alvin Weinberg admits that we may have only thirty to fifty years in which to adjust our energy consumption policies lest we impinge harmfully upon the world's atmospheric balance. Thermal pollution is in no sense a far-off, after-me-the-deluge disaster. It is a threat we can and must begin to avoid now. It is our duty, I believe, to re-arrange our energy consumption habits so that we do not hurtle blindly along until we trip the thermal pollution trigger and start a sequence of events, a tide, if you will, that our children may not be able to stem.

148

We have to understand that thermal pollution is a *different disaster*. We might indeed stumble into any number of catastrophes within our lifetimes and the lifetimes of our children. But these other disasters are not the absolutely *necessary* end to a succession of *current events*, the *unavoidable* outcome of our actions *today*. Every other problem we can think of can be prevented by taking steps that, compared to the adjustments we will have to make if we are not to melt the ice caps, are relatively simple. We have already demonstrated, for example, that we can avoid nuclear warfare. We have proved that we have the technological ability (if not the political will) to clean up air and water pollution without basically interfering with or changing the way we use oil, coal, and natural gas. We certainly know how we could, after some political and economic adjustments, provide enough food, right now, to lift all of the world's present population out of malnutrition and starvation.

But thermal pollution will be unavoidable unless we change at the most fundamental levels our competitive approach to national and international life. No filters on smoke stacks, no catalyzers on automobile engines, no ocean farms can stop thermal pollution if we go on accelerating our use of fossil fuels and nuclear energy. It is impossible to deny the mathematical, the physical, and the logical underpinnings of the argument. It is impossible to deny that thermal pollution is an unavoidable, an absolutely predictable, result of our present way of living, our present consumption patterns, our present acceleration of energy utilization.

Today we are confronted by a politically produced oil shortage. Even in the face of this shortage, politicians, economists, and business leaders are not saying that we should take the opportunity to learn to scale down permanently our thirst for energy. Not at all. They are saying that we should be conserving of energy until we have developed enough

149

new sources of fossil fuels and nuclear energy to reassert and reestablish our dizzying desires for and patterns of spiraling growth. Strenuous efforts to develop these sources of energy have already begun.

Any objective, nonideological view of the way the world's population is growing will also show that tens or possibly hundreds of billions of people will come to inhabit the earth. They will demand more goods and services—necessary goods as well as luxury goods—that will require the consumption of increasing amounts of energy. The laws of physics dictate that if this energy is produced from fossil fuels, nuclear power, geothermal sources, or huge solar energy satellites, the utilized energy will produce vast quantities of heat that must ultimately find their way to the poles and melt the ice caps.

If we are to avoid the calamity, there are several things we must do.

We must begin, today, to change our attitudes about growth. We are told by politicians, many economists, and virtually all business leaders that growth is a must. "Growth Is Not a Four Letter Word," an oil company ad in the *New York Times* said in 1972. "Only growth offers hope for solving many of the problems mankind faces. Without strong, sound, responsible growth, progress falters and halts. All that is wrong becomes worse, with even less hope for cure . . ."

The plea for unfettered growth is accepted as rational by many people; few stop to ask themselves just where this mania for growth is taking us. The truth is that we cannot afford to go on growing. Whether we like it or not, we live on a finite planet. Our world of limited temperature and limited space puts a limit on what we can do and the energy we can utilize. It does no good to pretend that wishful thinking or even new technology can change the very simple physical fact that limits are built into the planet on which we live. We will have to adjust to the realization that we cannot go on growing much longer, that the time will come, very soon, when man will have to use all his abilities just to be able to maintain his numbers in satisfactory fashion. We will have to come to grips with the realization that, national pride and corporate profits notwithstanding, we do not need billions of people and conspicuously consuming societies in order to enjoy good lives. Our entrapment in the world pattern of growth-now-growth-forever must be broken. And that break must start now, not when the melting of the ice caps has begun to send the oceans on their sweep into our cities and farm lands.

Billions of people will be born before we finally reach a no-growth plateau, and they must be given sufficient food, comfortable homes, a good living. Unavoidably, more and more energy will have to be used to support them. But that energy does not have to be thermally polluting energy. We should make a concerted effort, even as we strive to limit harmful growth, to make the transition as quickly as possible from the fossil fuels we are now using to the utilization of the earth's currently received solar energy. Since our leaders seem mesmerized by nuclear power, and since they seem determined to make it the main source of energy to meet our future needs, we should, as private citizens, take firm stands against the proliferation of fission and fusion power plants. It is time to stop cultivating the atom, time to start cultivating the sun!

Solar energy cells, giant windmills, giant turbines placed in the world's oceans are a few of the ways in which we could take advantage of solar energy falling on earth. The best and most extensive "converter" of solar energy will be, I believe, a system of open-ocean farms. By yielding vast amounts of vegetation, ocean farms can give us bountiful

151

amounts of nonthermally polluting fuels, huge amounts of fertilizers, plastics, and thousands of other products. Ocean farms can, in addition, supply us with man's single most important kind of energy: food.

Energy from ocean farms and other solar energy converters will be recognized to be very economical compared to energy from nuclear plants when the true costs of these plants—including the costs of nuclear terror—are included in the reckoning.

In sum, then, solar energy, used in combination with the oceans, can help us solve the three problems that will press us and our children most severely: thermal pollution, nuclear terror, and world famine.

But even the fullest use of solar energy cannot enable man to continue for long his ever accelerating spiral of growth. The thermal pollution catastrophe—in its awful prospect or in its hideous reality—will positively prevent man from ever utilizing the totality of the well-nigh boundless sources of energy on earth at a sustained rate higher than some 2.8×10^{21} Btus per year. This maximum rate is only about 10 percent more than the currently received solar input rate. Therefore, it matters not how plentiful fusion energy or fission energy or any other kind of energy may be, the earth's currently received solar energy will always make up at least 90 percent of the total that man will ever find permissible to use on this earth. But even the solar energy is limited. Therefore, though the utilization of ocean farms and other types of solar energy converters can temporarily palliate and postpone the present problems generated by man's spiraling growth, it cannot long resolve them. If man does not use this century and the next to develop a no-growth life style, the final and unavoidable hothouse catastrophe will terminate man's present civilization.

152

NOTES AND REFERENCES

CHAPTER 1: INTRODUCTION

PAGE

3. Energy is fundamental to all human life and natural processes. Nothing is a more important factor in tying together all the various aspects of science. What is it? The answer, shorn of technical niceties, is: the entity whose total magnitude remains constant but whose transformation or flow makes things happen. The force pushing a body in opposition to a resistance is delivering energy; a wheel rolling smoothly along a rail is carrying energy; the sun heating the ground and making the corn grow is supplying energy; the thinker's brain is transforming energy. See D. E. Roller and L. Nedelsky, "Energy," *Encyclopedia of Science and Technology*, McGraw-Hill, New York, 1971, vol. 4, pp. 686–688.

4. The reader may wonder at the use of the term "spiritual environment." I use it because a "threat" is by its very nature a psychological or spiritual entity. One who lives under threat necessarily suffers a spiritually impaired environment.

6. "Ten million will probably die [of starvation] this year—most of them children under 5 years old." *Newsweek*, Nov. 11, 1974, p. 56. One thousand calories of food energy is what a chemist or physicist would call 1000 Kilocalories of energy.

7. ". . . most of the world's abandoned oil reservoirs are still more than half full of crude petroleum . . ." See the National Petroleum Council's report on behalf of the U.S. oil industry: *U.S. Energy Outlook*, Dec. 1972, figure 12, p. 82.
   Regarding plans to produce liquid and gaseous fuels from coal, see the National Petroleum Council's report on behalf of the U.S. oil industry: *U.S. Energy Outlook*, Dec. 1972, pp. 354–375.

8. In reference to fossil fuels as sources of smog and other forms of air pollution, see the Report of the Environmental Pollution Panel of the President's Science Advisory Committee, *Restoring the Quality of Our Environment*, The White House, 1965, pp. 63, 96, 203–206. See also C. E. Billings and W. R. Matson, "Mercury

Emissions from Coal Combustion," *Science,* vol. 176, 1972, pp. 1232–1233.

9. To see that the swelling oceans will flood out most of our major cities if the ice caps melt, consider the fact that the level of the sea will then rise some 160 to 200 feet (M. M. Miller, "Glaciology," *Encyclopedia of Science and Technology,* McGraw-Hill, New York, 1971, p. 218.

10. See Chauncey Starr, "Energy and Power," *Scientific American,* vol. 225, Sept. 1971, pp. 36–49. How Dr. Starr derives his figure of 100 is not clear to me—I suspect that it is a bit high—but that is a small matter; I strongly agree with his main point.

   Regarding the assertions that a 1000-fold increase in man's fossil fuel and nuclear energy consumption rate will almost certainly bring on ice cap melting, and that a 10,000-fold increase will surely cause the ice caps to melt in a relatively few years, see M. I. Budyko, "The Effect of Solar Radiation Variations on the Climate of the Earth," *Tellus,* vol. 21, 1969, pp. 611–619; M. I. Budyko, "The Future Climate," *Transactions,* American Geophysical Union, vol. 53, 1972, pp. 868–874; W. D. Sellers, "A Global Climatic Model Based on the Energy Balance of the Earth-Atmosphere System," *Journal of Applied Meteorology,* vol. 8, 1969, pp. 392–400; W. D. Sellers, "A New Global Climatic Model," *Journal of Applied Meteorology,* vol. 12, 1973, pp. 241–54; Report of the Environmental Pollution Panel of the President's Science Advisory Committee, *Restoring the Quality of Our Environment,* The White House, 1965, p. 123. See also my own calculations, referenced in the footnote on page 14.

   The truth of the statement that "practically all of the energy powering every human activity . . . must ultimately warm the atmosphere . . ." proceeds from the following logic: (1) the energy cannot be destroyed (first law of thermodynamics); (2) the energy cannot be contained and must, in its totality, be continually exhibiting a net flow from higher temperature regions toward lower temperature regions (second law of thermodynamics); (3) the atmosphere is nearly (at least 90 percent) opaque to the transmission through it of long-wave radiation from the surface (W. D. Sellers, *Physical Climatology,* University of Chicago Press, 1965, p. 41); (4) if man could somehow refrigerate the atmosphere and send his waste energy directly into outer space by means of huge radiators placed *outside* the earth's atmosphere, the statement would be negated; (5) however, such a scheme would be prohibitively costly. For a reasonably fine-grained analysis of the limits placed by thermal pollution on man's ultimate population growth in the *absence* of all practical considerations such as excessive cost, see J. H. Fremlin, "How Many People Can the World Support?," *New Scientist*, no. 415, 1964, pp. 285–287.

11. See Wolf Häfele, "A Systems Approach to Energy," *American Scientist*, vol. 62, July–Aug. 1974, pp. 438–447. Note that Häfele was stating temperatures in Celsius degrees.

12. Regarding the ocean farm near San Clemente Island off the coast of California, see Chapter 8.

The table on page 156 indicates the probable abilities of the world's remaining supplies of fossil fuels plus nuclear energy to sustain man's ever accelerating energy consumption rate (average acceleration taken as constant at 5 percent per year) to levels well beyond the "thermal pollution flash point" (taken in this book as being 2.5×10^{20} Btu per year, or equivalently as being 10 percent of the rate at which solar energy falls on the surface of the earth).

CHAPTER 2: THE WATERS COME

13. Total area of Greenland glacier: 664,000 square miles; average thickness: 4,960 feet. Total area of Antarctica glacier: 4,650,000 square miles; average thickness: 5,460 feet. Combined volume: 5,380,000 cubic miles, equivalent to 4,860,000 cubic miles of meltwater. Total glacial ice of earth: 5,430,000 cubic miles. See E. J. Öpik, "Ice Ages," Chapter 10 of *The Earth and Its Atmosphere*, D. R. Bates, ed., Basic Books, New York, 1957, pp. 152–173.

For lowest temperatures of earth, see B. F. Howell, Jr., "Earth," *Encyclopedia of Science and Technology*, McGraw-Hill, New York, 1971, vol. 4, p. 364.

14. Regarding the fact that the total volume of the ice floating on the north polar sea is less than 0.2 of one percent of the volume of the earth's glacial ice, see E. J. Öpik, "Ice Ages," Chapter 10 of *The Earth and Its Atmosphere*, D. R. Bates, ed., Basic Books, New York, 1957, p. 162.

15. For confirmation that melting of the ice caps will raise the level of the sea relative to coastal lands by 160 to 200 feet, see (for example) M. M. Miller, "Glaciology," *Encyclopedia of Science and Technology*, McGraw-Hill, New York, 1971, vol. 6, p. 218.

Type of Energy Resource*	Estimated Quantity of Energy Remaining in Resource as of 1969 (Btu) (a)	Estimated 1969 Rate of Consumption of Resource ($\frac{Btu}{yr}$) (b)	Date of Resource Exhaustion at 5 Percent per Year Acceleration (Years AD)	Rate of Resource Consumption at Date of Exhaustion ($\frac{Btu}{yr}$)	Date at Which Rate of Resource Consumption Equals (or Would Equal) 10 Percent of Solar Energy Rate at Surface (Years AD)
Natural Gas	22×10^{18}	3.7×10^{16}	2040	1.3×10^{18}	2150
Coal & Lignite	340×10^{18}	6.7×10^{16}	2080	1.7×10^{19}	2130
Petroleum, Oil Shale, Tar Sands, & Bituminous Rocks	12×10^{21}	7.8×10^{16}	2148	6×10^{20}	2130
All Fossil Fuels Combined	12×10^{21}	1.9×10^{17}	2131	6×10^{20}	2110
Total Fissionable Fuels @ 100 Percent Burnup	12×10^{24}	starting at 1.3×10^{20} Btu/yr in 2100 AD	2269	5.9×10^{23}	2110
Total Deuterium for Fusion	1.5×10^{28}	same as above	2412	7.6×10^{26}	2110
Total Hydrogen for Fusion	10^{32}	same as above	2588	5×10^{30}	2110

* This table continues the discussion of the second note for page 12.

(a) Resource size as estimated in the U.S. Government documents *Energy R & D and National Progress—Findings and Conclusions of an Interdepartmental Study*, U.S. Government Printing Office, Washington, 1966, pp. 5–6; and A. B. Cambel, *Energy R & D and National Progress*, U.S. Government Printing Office, Washington, 1964, pp. 91–105. Resource size estimates given in these documents for the fossil fuels of the USA check closely with those given in the National Petroleum Council's report on behalf of the oil industry: *U.S. Energy Outlook*, Dec. 1972. (b) *UN Statistical Yearbook 1970*, table 12, p. 64.

PAGE

17. For details concerning the Bangladesh disaster of November 1971, see the *New York Times,* May 20, 1972.
In regard to the 1953 storm, see *Time,* Feb. 9, 1953; *Life,* Feb. 16, 1953; *Newsweek,* Feb. 9, 1953.

19. See C. Raghavan, "Bombay," *Encyclopaedia Britannica,* 1974, vol. 3, p. 15.

20. Following is my rough estimate (derived by eye from *Maps of the World,* Series 1142, U.S. Army Topographic Command and U.S. Naval Oceanographic Office, 1971–1972) of the areas of various states and countries (listed in alphabetical order) that will be flooded by the rising oceans if the ice caps melt:

State or Country	Area Loss (as Percent of Present Area)
Afghanistan	0
Alabama, USA	40
Alaska, USA	20
Albania	10
Algeria	2
Angola	5
Argentina	20
Arizona, USA	0.5
Arkansas, USA	30
Australia	30
Austria	10
Bangladesh	50+
Belgium	20
Bolivia	2
Botswana	0
Brazil	20
British Honduras	20
Bulgaria	10
Burma	30
California, USA	10
Cambodia	40
Cameroon	2
Central African Republic	0
Chad	2
Chile	10

State or Country	Area Loss (as Percent of Present Area)
Colombia	20
Colorado, USA	0
Congo	5
Connecticut, USA	40
Cuba	40
Czechoslovakia	2
Dahomey	10
Delaware, USA	50+
Denmark	50+
Dominican Republic	20
East Germany	40
East Russia (90°E to 180°E)	30
Ecuador	20
Egypt	20
El Salvador	30
England	30
Equatorial Guinea	10
Ethiopia	2
Finland	40
Florida, USA	100
France	20
French Guiana	40
French Territory of Afars and Issas	10
Gabon	20
Georgia, USA	40
Ghana	30
Greece	10
Guatemala	20
Guiana	30
Guinea	10
Haiti	20
Hawaii, USA	10
Holland	100
Honduras	10
Hungary	30
Iceland	10
Idaho, USA	0
Illinois, USA	0
India	20

NOTES AND REFERENCES

State or Country	Area Loss (as Percent of Present Area)
Indiana, USA	0
Indonesia	40
Iowa, USA	0
Iran	1
Iraq	30
Ireland	30
Israel	20
Italy	20
Ivory Coast	20
Japan	20
Jordan	1
Kansas, USA	0
Kentucky, USA	0
Kenya	5
Kuwait	40
Lesotho	0
Liberia	20
Libya	10
Louisiana, USA	100
Luxembourg	20
Maine, USA	20
Malagasy Republic	10
Malawi	0
Malay Peninsula	40
Mali	2
Maryland, USA	30
Massachusetts, USA	20
Mauritania	20
Mexico	10
Michigan, USA	0
Minnesota, USA	0
Mississippi, USA	100
Missouri, USA	0
Mongolia	0
Montana, USA	0
Mozambique	30
Nebraska, USA	0
Nevada, USA	0
New Hampshire, USA	20

159

State or Country	Area Loss (as Percent of Present Area)
New Jersey, USA	40
New Mexico, USA	0
New York, USA	10
Nicaragua	30
Niger	0
Nigeria	10
North Carolina, USA	20
North Dakota, USA	0
North Vietnam	20
Northern Ireland	30
Norway	10
Ohio, USA	0
Oklahoma, USA	0
Oregon, USA	10
Pakistan	30
Panama	20
Paraguay	30
Pennsylvania, USA	10
People's Republic of China	10
Peru	20
Philippines	30
Poland	40
Portugal	20
Portuguese Guinea	50+
Puerto Rico	20
Rhode Island, USA	100
Saudi Arabia	7
Scotland	20
Senegal	50+
Sierra Leone	30
Somalia	10
South Africa	2
South Carolina, USA	40
South Dakota, USA	0
South Vietnam	20
Southern Rhodesia	0
Southwest Africa	2
Spain	2

	Area Loss (as Percent of Present Area)
State or Country	
Spanish Sahara	20
Sudan	0
Sumatra	40
Surinam	20
Swaziland	20
Sweden	20
Switzerland	0
Syria	0
Taiwan	20
Tanzania	2
Tennessee, USA	10
Texas, USA	20
Thailand	30
Togo	20
Turkey	5
Uganda	0
Utah, USA	0
Venezuela	20
Vermont, USA	10
Virginia, USA	30
Wales	20
Washington State, USA	20
West Germany	20
West Russia (20°E–90°E)	40
West Virginia, USA	0
Wisconsin, USA	0
Wyoming, USA	0
Yugoslavia	10
Zaire	0
Zambia	0

PAGE

22. The table of cities is derived from the *New Cosmopolitan World Atlas,* Rand McNally, New York, 1965, pp. 198 and 216.

Clearly, the urban populations of the world live mostly on coastlines below 100 feet of altitude. Indeed, of the 245 million people represented by the table, some 58 percent (144 million) live below 100 feet, 62 percent (155 million) live below 160 feet, and 64 percent (159 million) live below 200 feet of altitude.

24. In regard to "triggering" an ice cap melting disaster, see pages 36-40.

CHAPTER 3: HOTHOUSE EARTH

27. That the natural (i.e., non-man-made) heat of the lands, ocean, and atmosphere of the earth comes almost entirely from the sun is seen in the following table (derived from W. D. Sellers' *Physical Climatology*, University of Chicago Press, 1965, p. 12):

Source	Btu per Year	Percent of Solar
Solar energy received at surface of earth	2.5×10^{21}	100
Natural geothermal heat	9.5×10^{17}	3.8×10^{-2}
Man's energy consumption	2.0×10^{17}	8×10^{-3}
Long-wave radiation from full moon	1.6×10^{17}	6.4×10^{-3}
U.S. energy consumption	7.0×10^{16}	2.8×10^{-3}
Solar radiation reflected from full moon	5.3×10^{16}	2.1×10^{-3}
Solar tides in atmosphere	5.3×10^{16}	2.1×10^{-3}
Lightning discharges	3.2×10^{15}	1.3×10^{-4}
Magnetic storms	1.1×10^{15}	4.2×10^{-5}
Cosmic rays	4.8×10^{14}	1.9×10^{-5}
Meteorites	3.2×10^{14}	1.3×10^{-5}
Starlight	2.1×10^{14}	8.5×10^{-6}
Lunar tides in atmosphere	1.6×10^{14}	6.4×10^{-6}
All nonsolar combined	1.5×10^{18}	5.97×10^{-2}

28. The term "hothouse effect" used in this book is often replaced by the term "greenhouse effect." See L. D. Kaplan, "Greenhouse Effect, Terrestrial," *Encyclopedia of Science and Technology*, McGraw-Hill, New York, 1971, vol. 6, p. 305. The first description of this effect, so far as I am aware, was given in J. Tyndall, "On Radiation Through the Earth's Atmosphere," *Philosophical Magazine*, 4th Series, 25, 1863, pp. 200–206.

29. That the net flow of heat is always from higher to lower temperature regions, one of the most basic laws of nature, is called "the

second law of thermodynamics." See G. A. Hawkins, "Thermodynamic Principles," *Encyclopedia of Science and Technology,* McGraw-Hill, New York, 1966, p. 567. (The "first law" of thermodynamics states that energy cannot be created or destroyed, though it can in certain ways be transformed from one kind to another. Ibid., p. 566.)

30. Regarding the earth being devoid of ice at both poles for most of the last 500-million-year period, see *Report of the Study of Man's Impact on Climate,* MIT Press, Cambridge, Mass., 1971, pp. 9 and 29.

31. For the temperature history of the earth before and during the Quaternary Age, see *Report of the Study of Man's Impact on Climate,* MIT Press, Cambridge, Mass., 1971, p. 31.

32. A positive or negative feedback mechanism is a connection from the output (effect) of a machine around to its input (cause), so that the output either reinforces or opposes the input. See H. F. Klock, "Feedback Circuit," *Encyclopedia of Science and Technology,* McGraw-Hill, New York, 1971, vol. 5, pp. 214–216.

    For the excellent phrase "climate machine," plus many valuable insights into the subject of this book, I am indebted to the article by J. O. Fletcher, "Polar Ice and the Global Climate Machine," *Bulletin of the Atomic Scientists,* Dec. 1970, pp. 40–47.

    The "ice cap reflectivity mechanism" is explained in many books and articles; see, for example, Sverdrup, Johnson, and Fleming, *The Oceans,* Prentice-Hall, Englewood Cliffs, N.J., 1942, pp. 113–114.

    The "carbon dioxide–ocean temperature mechanism" is discussed in T. C. Chamberlin, "An Attempt to Frame a Working Hypothesis of the Cause of Glacial Periods on an Atmospheric Basis," *Journal of Geology,* vol. 7, 1899, pp. 574–575. See also G. N. Plass, "The Carbon Dioxide Theory of Climatic Change," *Tellus,* vol. 8, 1956, pp. 140–154.

    The "water vapor–ocean temperature" feedback mechanism is described in considerable depth in A. P. Ingersoll, "The Runaway Greenhouse: A History of Water on Venus," *Journal of the Atmospheric Sciences,* vol. 26, pp. 1191–1198.

33. Negative feedback mechanisms can be seen operating nearly everywhere in nature. For example, the pain produced when a child puts his hand forward into a fire is a negative feedback mechanism, for its effect (the abrupt withdrawal of the hand) *opposes* the motion (the putting forth of the hand into the fire)

which caused it (the pain). Another example: the flow of air (wind) from a high pressure zone toward a low pressure zone is a negative feedback mechanism, for this flow of air causes the pressure to fall in the high pressure zone and to rise in the low pressure zone, whence its effect is to *reduce or oppose* the atmospheric conditions (the existence of zones having different pressures) which cause it (the flow of the air).

Indeed, the stability of the physical and social environments on earth, wherever such stability exists, is generally the result of the action of some negative feedback mechanisms. But wherever an instability does occur, as in the development of a storm in the atmosphere or a revolution in the social structure, one can generally find a *positive* feedback mechanism at work.

34. For a general description of how oscillatory systems ("oscillators") work, see H. F. Klock, "Feedback Circuit," *Encyclopedia of Science and Technology*, McGraw-Hill, New York, 1971, vol. 5, pp. 214–216. It will be observed from this reference—see p. 215—that when the strength or "gain" of the positive feedback mechanism operating in an oscillator is reduced, a critical level of gain may be reached below which the oscillator stops its free oscillations and becomes simply a powerful (or "high gain") amplifier. If the gain of the positive feedback mechanism is then reduced still farther, the amplifier becomes more and more stable —i.e., more and more "low gain" in its operation.

35. In reference to the slowness which would probably characterize changes in the earth's climate if it were "left to itself" without the addition of man's interfering activities, see W. W. Kellogg, "Long-Range Influences of Mankind on the Climate," National Center for Atmospheric Research, Boulder, Colo., 1974, p. 14.

36. Using the *UN Statistical Yearbook, 1970*, Table 12, p. 64, we can derive the following table of values:

Type of Energy Resource	Annual Percentage Acceleration in Resource Consumption
Natural Gas	8.25
Petroleum	7.00
Coal & Lignite	2.09
All Terrestrial Resources	4.93

Using man's 1969 energy consumption rate of 1.87×10^{17} Btu/yr (ibid.) and a projection of man's consumption of terrestrial energy

164

resources at constant acceleration levels chosen to bracket the above value of 4.93 percent per year, we can derive the following chart:

	Time When Man's Global Energy Consumption Rate Reaches Indicated Level				
Acceleration Values (Percent per Year)	1/1000th Solar $(2.5 \times 10^{18}$ Btu/Yr)	1/100th Solar $(2.5 \times 10^{19}$ Btu/Yr)	1/10th Solar $(2.5 \times 10^{20}$ Btu/Yr)	Solar Rate $(2.5 \times 10^{21}$ Btu/Yr)	10 x Solar Rate $(2.5 \times 10^{22}$ Btu/Yr)
4	2034 AD	2091 AD	2149 AD	2207 AD	2264 AD
6	2012 AD	2051 AD	2089 AD	2127 AD	2166 AD

37. As to the definition of "pollution," see the Report of the Environmental Pollution Panel of the President's Science Advisory Committee, *Restoring the Quality of Our Environment*, The White House, 1965, p. 1.

38. See M. I. Budyko, "The Future Climate," *Transactions*, American Geophysical Union, vol. 53, 1972, p. 871. See also, M. I. Budyko, *Climate and Life*, Academic Press, New York, 1974, pp. 306–308.

39. See W. D. Sellers, "A Global Climatic Model Based on the Energy Balance of the Earth–Atmosphere System," *Journal of Applied Meteorology*, vol. 8, 1969, pp. 392–400; see also W. D. Sellers, "A New Global Climatic Model," *Journal of Applied Meteorology*, vol. 12, 1973, pp. 241–254.

For a statement of observed variations in sea ice coverage at the North Pole, see the Report of the Environmental Pollution Panel of the President's Science Advisory Committee, *Restoring the Quality of Our Environment*, The White House, 1965, pp. 123–24; see also W. D. Sellers, "A New Global Climatic Model," *Journal of Applied Meteorology*, vol. 12, 1973, p. 251, and J. O. Fletcher, "Polar Ice and the Global Climate Machine," *Bulletin of the Atomic Scientists*, Dec. 1970, pp. 40–47.

CHAPTER 4:   THE WORLD FOOD CRISIS

42. T. Malthus, *An Essay on the Principle of Population*, first published in 1798; available in G. Hardin, *Population, Evolution, and Birth Control*, Freeman & Co., San Francisco, 1964, pp. 1–20.

42. P. Handler, "On the State of Man," An Address Presented to the Annual Convocation of Markle Scholars at the Homestead on Sept. 29, 1974, National Academy of Sciences, p. 6.

Regarding 33 to 50 percent of the world's people being hungry, see N. E. Borlaug, "Civilization's Future," *Science and Public Affairs,* vol. 29, 1973, p. 9.

"More than 5 million children under 5 years of age are starving to death each year at the present time." *Newsweek,* Nov. 11, 1974, p. 56.

In regard to Handler's statements, see P. Handler, "On the State of Man," An Address Presented to the Annual Convocation of Markle Scholars at the Homestead on Sept. 29, 1974, National Academy of Sciences, pp. 6, 11.

43. The moral (or other) obligation of the well-off to feed the poor is a knotty issue. The position taken by the author—a position not argued in this book because to do so would be too extensive and too diversionary, but nevertheless a position which *underlies* much of the argument of this book—is that *both* the well-off and the poor *jointly* bear *two* closely coupled obligations of a moral nature: (1) the "total society of man" should *somehow* try seriously to ensure that all who are permitted to be born into this world get adequate food and other care, and (2) this same "total society of man" should *somehow* figure out how to decide and determine who and how many are to be born into this world. Clearly, the two "somehows" italicized above involve many questions which reasonable people may differ about; nevertheless, the establishment of *provisional answers* to the questions raised by these "somehows" is an urgent task for world politicians.

Borlaug's estimates: 8 pounds of grain consumed for each pound of beef supplied, and 1800 pounds of grain consumed per year by the average American, as compared to 400 pounds consumed by the average person in the underdeveloped nations, can be found in the document "U.S. and World Food Situation: Hearings Before the Subcommittee on Agricultural Production . . . , U.S. Senate, Oct. 17 and 18, 1973," U.S. Government Printing Office, Washington, 1974, p. 12. See also N. S. Scrimshaw, "The Worldwide Confrontation of Population and Food Supply," *Technology Review,* Dec. 1974, pp. 12–19.

L. R. Brown, *By Bread Alone,* Praeger, New York, 1974, p. 44.

44. That there are almost 100 million more people to feed each year reflects the net growth rate of the population (more than 2 percent

PAGE

per year) applied to the total human population on earth (almost exactly 4 billion persons). See D. Nortman, "Quantity vs. Quality of Life," *Bulletin of the Atomic Scientists,* June 1974, p. 13–16.

45. In reference to human population numbers over the centuries and millennia, see M. King Hubbert, *Energy Resources,* Publication 1000-D of the National Academy of Sciences—National Research Council, Washington, 1962, 15–21.

For gross birth rates and death rates, see D. Nortman, "Quantity vs. Quality of Life," *Bulletin of the Atomic Scientists,* June 1974, p. 15.

46. If man's population had started 30,000 years ago at a level of one million persons, say, and if it had then grown at an acceleration of 2.0 percent per year for 30,000 years, his numbers at present would be some $10^6 \mathrm{x} e^{0.020 \mathrm{x} 30,000} = 10^6 \mathrm{x} e^{600} = 10^{267}$ persons. Taking the average volume of a person as about 2 cubic feet, say, the volume of all humanity would then be 2×10^{267} cubic feet. The volume of the "presently visible universe" is about 2×10^{80} cubic feet (A. Sandage, "Galaxy, External," *Encyclopedia of Science and Technology,* McGraw-Hill, New York, 1971, vol. 6, p. 12), so the volume of humanity would fill not just one but some 10^{187} visible universes.

For populations of today's China, United States, and Europe, see D. Nortman, "Quantity vs. Quality of Life," *Bulletin of the Atomic Scientists,* June 1974, p. 15.

As a more or less typical statement of the position of the "population optimists," see H. Kahn and B. Bruce-Biggs, *Things to Come,* Macmillan, New York, 1972, p. 23. For the opposite view, see K. Davis, "The Changing Balance of Births and Deaths," Chapter 1 of *Are Our Descendants Doomed?,* Brown and Hutchings, eds., Viking, New York, 1970, pp. 17–19.

47. For the story of the population growth forecasts by the UN in 1954, and for much other good information as well, see H. Brown, J. Bonner, and J. Weir, *The Next Hundred Years,* Viking, New York, 1957.

For Harrison Brown's statement, see *The Next Ninety Years,* California Institute of Technology, Pasadena, California, 1967, p. 8.

48. In respect to continued population increases to be expected soon after reaching a replacement rate characteristic of zero long-term population growth, see K. Davis, "The Changing Balance

of Births and Deaths," Chapter 1 of *Are Our Descendants Doomed?*, Brown and Hutchings, eds., Viking, New York, 1970, pp. 30–31.

For the fact that population increase is caused not by *unwanted* but rather by *wanted* children, see K. Davis, "The Changing Balance of Births and Deaths," Chapter 1 of *Are Our Descendants Doomed?*, Brown & Hutchings, eds., Viking, New York, 1970, p. 24.

49. For the speech by Pope Paul VI, see the *Los Angeles Times,* Nov. 10, 1974, p. 1.

50. Regarding the arable land acreage of the world, see *Newsweek,* Nov. 11, 1974, pp. 56–68. See also the President's Science Advisory Committee report, *The World Food Problem,* The White House, 1967, vol. II, pp. 422–435.

51. Regarding the irrigation wells in India and Pakistan, see L. Brown, *By Bread Alone,* Praeger, New York, 1974, pp. 98–99.

 In reference to the West Texas situation, see L. Brown *By Bread Alone,* Praeger, New York, 1974, p. 103.

 For L. Brown's statement, see his book, *By Bread Alone,* Praeger, New York, 1974, p. 9.

 Regarding use of desalted ocean water for irrigation, see G. Borgstrom, *The Food and People Dilemma,* Duxbury Press, North Scituate, Mass., 1973, p. 41.

52. For statements about the Green Revolution, see "U.S. and World Food Situation: Hearings before the Subcommittee on Agricultural Production . . . , U.S. Senate, Oct. 17 and 18, 1973," U.S. Government Printing Office, Washington, 1974; see also *Newsweek,* Nov. 11, 1974, pp. 62 and 67, and N. S. Scrimshaw, "The World-wide Confrontation of Population and Food Supply," *Technology Review,* Dec. 1974, pp. 12–19.

53. For the statement by G. Barraclough, professor of history at Brandeis University, see his article, "The World Crash," *New York Review of Books,* vol. 21, Jan. 23, 1975, pp. 27–28.

 For Brown's statement, see L. Brown, *By Bread Alone,* Praeger, New York, 1974, p. 77.

 See the President's Science Advisory Committee report, "The World Food Problem," The White House, May, 1976, vol. II, p. 480.

 L. Brown's statements can be found in his book *By Bread Alone,* Praeger, New York, 1974, p. 79 and pp. 86–87.

PAGE

54. See G. Borgstrom, *The Food and People Dilemma*, Duxbury Press, North Scituate, Mass., 1973, p. 39.
55. See L. Brown, *By Bread Alone*, Praeger, New York, 1974, p. 87.

CHAPTER 5: STOKING HOTHOUSE EARTH: PART 1

58. Dr. A. B. Cambel's statement may be found in *National Goals Symposium, Part 1*, Hearings Before the Committee on Interior and Insular Affairs, U.S. Senate, Oct. 20, 1971, p. 3.
See Federal Power Commission report, *The 1970 National Power Survey*, U.S. Government Printing Office, Washington, 1971.
59. In reference to the per capita income and per capita energy use figures for various countries, see W. Häfele, "A Systems Approach to Energy," *American Scientist*, vol. 62, July–Aug. 1974, p. 439.
60. For the facts concerning the growth in per capita energy consumption in the USA in the century from 1850 to 1950, see P.C. Putnam, *Energy in the Future*, Van Nostrand, New York, 1953, p. 87.
61. For some fairly recent values of some energy use efficiencies in the USA, see P. C. Putnam, *Energy in the Future*, Van Nostrand, New York, 1953, Table 4–24N, pp. 390–395.
62. The unattributed quotation is from a private communication.
63. The Borlaug quotation is from "U.S. and World Food Situation: Hearings Before the Subcommittee on Agricultural Production ..., U.S. Senate, Oct. 17 and 18, 1973," U.S. Government Printing Office, Washington, 1974, p. 12.
64. The unattributed quotation is from a private communication.
A rough estimate of the energy required to produce and operate an automobile throughout its life may be obtained as follows: In 1970 the USA consumed about 6.8×10^{16} Btus of energy (*U.S. Energy Outlook*, National Petroleum Council, 1972, p. 37) and had a gross national product of about \$0.977 trillion (*World Almanac, 1974*, Newspaper Enterprise Association, New York, 1974, p. 85). Therefore, each U.S. dollar in 1970 represented roughly 7×10^4 Btu. To make a typical \$4000 automobile in 1970 required the direct expenditure of roughly 3.7×10^4 kwh = 1.3×10^8 Btu (R. S. Berry and M. F. Fels, in *Science and Public Affairs*, Dec. 1973, p. 13). Hence the energy required to fabricate the car represented a cost of roughly 1.3×10^8 Btu ÷ 7×10^4 Btu per dollar = \$1900. The total price of the car, naturally, includes the cost represented by all this energy plus \$2100 for the pro rata cost to that car of the buildings, transportation, sales efforts,

profits, taxes, and all the rest required to deliver the car to the customer; therefore, the energy represented by all the latter items is about 1.4×10^8 Btu—some 10 percent more than that required to make the car itself. Next, the energy required to drive that car over its life span (taken as roughly 10^5 miles) at some 14 miles per gallon of gasoline (*U.S. Statistical Abstract, 1972,* table 898, p. 549) and 37 kwh per gallon of gasoline, (*Handbook of Chemistry & Physics,* 36th ed., Chemical Rubber Publishing Co., Cleveland, Ohio, 1954, p. 1757) is 10^5 miles $\times \dfrac{1 \text{ gal}}{14 \text{ mile}} \times \dfrac{37 \text{ kwh}}{\text{gal}} =$ 2.6×10^5 kwh $= 9.3 \times 10^8$ Btu. Thus the total energy represented by the car is roughly 1.2×10^9 Btu, of which about 77 percent is for operating the car over its lifespan, about 11 percent is for fabricating the car, and about 12 percent is for making all the factories, sales rooms, garages, and so forth, required for delivering the car as a product to the market. (For a similar but more extensive analysis, see E. Hirst, "Energy Consumption for Transportation in the U.S.," Oak Ridge National Laboratory, Report No. ORNL–NSF–EP–15, March 1972, pp. 20–24 and 32–34.)

67. One of the main points of this book—that the ice caps will surely melt (and later the oceans will surely boil) if man does not terminate the acceleration of his energy consumption spiral—is analogous to the principal message of the now famous study, *The Limits to Growth* (D. H. Meadows, et al., Universe Books, New York, 1972). That book was strange, however, in that its major conclusion (as foreshadowed in its title) was essentially *correct* at the same time that its argumentative details were *inadequate.* Many of its severest critics attacked its details without acknowledging the validity of its general conclusion. The present book, I hope, differs from the "Limits" book in that the validity of this book's message is almost completely *independent* of the many minor details of the reasoning. No complex computerized model is required to prove and drive home the truth of this book's message, no great appeal is made to questionable climatological factors, no reliance is placed on obscure behavior patterns of individuals or societies, no argumentation is derived from poorly understood concepts such as "quality of life." I therefore believe that no expert and no sensible person can contradict my point: the ice caps will surely melt (and later the oceans will surely boil) if man does not terminate the acceleration of his energy consumption spiral.

CHAPTER 6:   STOKING HOTHOUSE EARTH: PART 2

PAGE

69. That there is "more than enough oil out there to meet our current needs" can be verified by reference to the second note for page 12 and the table on page 156.

72. For the USA's proved oil reserves, see National Petroleum Council report, *U.S. Energy Outlook*, Dec. 1972, table 38, page 72. For the total estimated potential petroleum resources of the whole world, see A. B. Cambel, *Energy R & D and National Progress*, U.S. Government Printing Office, Washington, 1964, p. 105.

To verify the statement that 60 to 80 percent of the original oil is still left in oil wells that have been or are about to be shut down, see the report by the oil industry's National Petroleum Council, *U.S. Energy Outlook*, Dec. 1972, fig. 12, p. 82.

In reference to methods for extracting more oil from old reservoirs, see E. G. Dahlgren, "Petroleum Secondary Recovery," *Encyclopedia of Science and Technology*, McGraw-Hill, New York, 1971, vol. 10, pp. 93–95.

For a description of the quantities of oil that can be recovered from existing oil wells under various assumed economic conditions, see the report of the oil industry's National Petroleum Council, *U.S. Energy Outlook*, Dec. 1972, pp. 62–133. See also A. B. Cambel, *Energy R & D and National Progress*, U.S. Government Printing Office, Washington, 1964, p. 154.

74. Regarding the origin of oil, see J. M. Hunt, "Petroleum, Origin of," *Encyclopedia of Science and Technology*, McGraw-Hill, New York, 1971, vol. 10, pp. 66–67. For a discussion of the origins of oil sands and oil shale, see the articles by I. A. Breger, "Oil Sand" and "Oil Shale," *Encyclopedia of Science and Technology*, McGraw-Hill, New York, 1971, vol. 9, pp. 350–354.

75. For a discussion of oil shale extraction technologies, see I. A. Breger, "Oil Shale," *Encyclopedia of Science and Technology*, McGraw-Hill, New York, 1971, vol. 9, pp. 351–354.

76. A pessimistic viewpoint regarding oil shale abundance can be found in M. K. Hubbert, *Energy Resources*, Publication 1000-D of the National Academy of Sciences–National Research Council, Washington, 1962, p. 89. A more optimistic estimate can be found in A. B. Cambel, *Energy R & D and National Progress*, U.S. Government Printing Office, Washington, 1964, last full paragraph of p. 104.

171

76. The German capability for producing liquid fuel from coal is given in J. H. Field, "Bergius Process," *Encyclopedia of Science and Technology,* McGraw-Hill, New York, 1971, vol. 2, p. 165.

77. Information regarding costs and uses of methanol can be found in T. B. Reed and R. M. Lerner, "Methanol: A Versatile Fuel for Immediate Use," *Science,* vol. 182, 1973, pp. 1299–1304. See also E. E. Wigg, "Methanol as a Gasoline Extender: A Critique," *Science,* vol. 186, 1974, pp. 785–790.

For a reasonably sound and readable discussion of geothermal energy—its potentials and some of its problems—see C. P. Gilmore, "Hot New Prospects for Power from the Earth," *Popular Science,* vol. 201, no. 2, Aug. 1972, p. 56.

78. In regard to the total rate at which geothermal energy flows up out of the earth's land surfaces and ocean bottoms, see S. P. Clark, Jr., "Earth, Heat Flow in," *Encyclopedia of Science and Technology,* McGraw-Hill, New York, 1971, vol. 4, pp. 370–373. See also the note for page 27.

79. In regard to environmental effects to be expected from California's geothermal sources, see M. Goldsmith, "Geothermal Resources in California: Potentials and Problems," Report No. 2, Environmental Quality Laboratory, California Institute of Technology, Pasadena, 1971.

For the possibilities of extracting geothermal energy from dry sources, see "Geothermal Energy Resources and Research," Hearings of the Interior and Insular Affairs Committee, U.S. Senate, June 15 and 22, 1972, pp. 269–307.

80. In respect to searching for geothermal outcroppings by use of satellites, see D. L. Peck, "Assessment of Geothermal Energy Resources," Federal Council for Science and Technology, 1972, p. 42 (included in reference cited in note for page 79, above).

For the 1981 power output expected from the geothermal field called The Geysers, see "The Potential for Geothermal Power," *Business Week,* March 17, 1973, p. 74.

The world-wide search for geothermal sources is briefly set forth in the second reference cited for page 77 and the first reference for page 80.

81. In reference to the facts given about energy from nuclear fission, see articles by L. J. Koch, "Reactor, Nuclear," vol. 11, pp. 375–383; H. S. Isbin, "Reactor, Nuclear (Classification)," vol. 11, pp. 383–386; B. I. Spinrad, "Reactor Physics," vol. 11, pp. 389–395; J. A. Lane, "Nuclear Fuels," vol. 9, pp. 207–211;

NOTES AND REFERENCES

PAGE

    S. Lawroski and R. C. Vogel, "Nuclear Fuels Reprocessing," vol. 9, pp. 211–213; J. A. Lane,, "Nuclear Power," vol. 9, pp. 220–224; *Encyclopedia of Science and Technology*, McGraw-Hill, New York, 1971.

81. In regard to the nuclear power plants which the government is planning, see D. Burnham, "Congress Faces Key Decisions on Nuclear Reactors," *New York Times*, March 30, 1975, p. 26.

83. For an estimate of the total energy potentially available from fissionable fuels, see the second note for page 12 and the table on page 156.

    For a good discussion of fusion reactor facts and possibilities, see R. F. Post, "Fusion, Nuclear," *Encyclopedia of Science and Technology*, McGraw-Hill, New York, 1971, vol. 5, pp. 630–633. See also R. F. Post and F. L. Ribe, "Fusion Reactors as Future Energy Sources," *Science*, 1974, vol. 186, pp. 397–407. In regard to laser-induced fusion, see W. D. Metz, "Laser Fusion: A New Approach to Nuclear Power," *Science*, vol. 177, 1972, pp. 1180–1182, and "Laser Fusion: One Milestone Passed—Millions More to Go," *Science*, vol. 186, 1974, pp. 1193–1195.

84. To find the gross fusion energy value represented by the deuterium present in one gallon of sea water, we proceed as follows: In each gallon of ocean water there are some 8.3 pounds of water or about 0.93 pounds of hydrogen nuclei. There is about one nucleus of deuterium for every 6700 nuclei of hydrogen (J. Bigeleisen, "Deuterium," *Encyclopedia of Science and Technology*, McGraw-Hill, New York, 1971, vol. 4, pp. 102–103), and each deuterium nucleus weighs roughly twice as much as each hydrogen nucleus, so in one gallon of water there are some 2.8×10^{-4} pounds or 1.3×10^{-1} grams of deuterium. But each gram of deuterium potentially represents about 10^5 kwh of fusion energy release, (R. F. Post, "Fusion, Nuclear," *Encyclopedia of Science and Technology*, McGraw-Hill, New York, 1971, vol. 5, p. 631), so each gallon of sea water represents about 1.3×10^4 kwh or 4.3×10^7 Btu of this kind of energy. Since the complete combustion of one gallon of gasoline represents an energy release of some 37 kwh $= 1.3 \times 10^5$ Btu (*Handbook of Chemistry & Physics*, 36th ed., Chemical Rubber Publishing Co., Cleveland, Ohio, 1954, p. 1757), we can say that each gallon of sea water contains a level of "deuterium fusion energy" equal to the combustion energy of 340 gallons of gasoline. (Note: this calculation considers gross energy delivered to the environment, not the part of that energy which can be made useful to man.)

84. For the energy resource comparisons given, see the second of the notes for page 12 and the table on page 156.

The "natural laws of human nature and human societies" are slowly emerging, I believe, as the "science of human evolution" continues gradually to be developed. Fundamentally, I conceive of this science as postulating that human individuals and societies evolve in such a way as to improve the ability of the human species to survive in the face of challenges mounted in the past by the environment. The laws governing this process are what I mean by the "natural laws of human nature and human societies."

85. For a good discussion of the turbulent controversies over the question of removing sulfur from the stack gases of coal-fired power stations, see R. Stuart, "Utilities' Compliance with Clean-Air Standards Proceeds Slowly," *New York Times,* April 14, 1975, pp. 49, 52.

86. Regarding pollution from geothermal sources, see M. Goldsmith, "Geothermal Resources in California: Potentials and Problems," Report No. 2, Environmental Quality Laboratory, California Institute of Technology, Pasadena, 1971.

Many of the atomic nuclei present in the radioactive wastes from fission plants have the habit or property of "exploding" every so often. When a mass of wastes has many such explosions occurring in it during each second of time, a large number of more or less energetic and hence damaging nuclear fragments are being shot forth from the material, and this is what makes such wastes dangerous. When each radioactive nucleus explodes, it usually becomes another type of radioactive nucleus, and this then produces another explosion after some period of time. When an adequate number of the original nuclei have finally been transformed through a sequence of explosions into nonradioactive (stable) types of nuclei, the rate of emission of damaging fragments from the wastes will have dropped to a safe level and the wastes are then said to have "cooled off." This process can take thousands of years if the original levels of radioactivity are high—see R. D. Evans, "Radioactivity," *Encyclopedia of Science and Technology,* McGraw-Hill, New York, 1971, vol. 11, pp. 286–298; see also K. Z. Morgan, "Radioactive Waste Disposal," ibid., vol. 11, pp. 284–286.

For a discussion of the quantity of nuclear waste, see K. Z. Morgan, "Radioactive Waste Disposal," *Encyclopedia of Science and Technology,* McGraw-Hill, New York, 1971, vol. 11, pp. 284–286.

NOTES AND REFERENCES

PAGE

86. In regard to using a rocket ship to dispose of radioactive wastes, see F. K. Pittman, "Management of Commercial High-Level Wastes," Waste Management and Transportation Division, U.S. Atomic Energy Commission, 1972, pp. 2 and 4. See also A. M. Weinberg, "Social Institutions and Nuclear Energy," *Science*, vol. 177, 1972, pp. 27–34.

87. Salt mines as possible disposal sites are discussed in K. Z. Morgan, "Radioactive Waste Disposal," *Encyclopedia of Science and Technology*, McGraw-Hill, New York, 1971, vol. 11, p. 286. See also A. M. Weinberg, "Social Institutions and Nuclear Energy," *Science*, vol. 177, 1972, pp. 27–34.

    For remarks by T. B. Taylor, see the *Los Angeles Times*, Dec. 8, 1974, part I, p. 13.

88. Descriptions of the nuclear accidents mentioned can be found in S. Novick, *The Careless Atom*, Houghton Mifflin, Boston, 1969.

90. Regarding the fact that plutonium is "the most poisonous existing substance," see J. G. Speth, A. R. Tamplin, and T. B. Cochran, "Plutonium Recycle: The Fateful Step," *Bulletin of the Atomic Scientists*, Nov. 1974, p. 15. See also J. P. Holdren, "Hazards of the Nuclear Fuel Cycle," *Bulletin of the Atomic Scientists*, Oct. 1974, pp. 14–23.

    For a disturbing description of the apparent lack of *official* concern shown by the AEC in regard to the safety of nuclear reactors, see R. Gillette, "Nuclear Reactor Safety: At the AEC the Way of the Dissenter Is Hard," *Science*, vol. 176, 1972, pp. 492–498.

91. See R. Gillette, "Nuclear Safety (I): The Roots of Dissent," *Science*, vol. 177, 1972, p. 771.

    The MIT study is available as: N. C. Rasmussen, "Reactor Safety Study—An Assessment of Accident Risks in U.S. Commercial Nuclear Power Plants," National Technical Information Service, Dept. of Commerce, Springfield, Va. 22151.

92. Philip Handler, "On the State of Man," An Address Presented to the Annual Convocation of Markle Scholars at the Homestead on Sept. 29, 1974, National Academy of Sciences, p. 13.

93. For the position of some proponents of fusion energy plants, see R. F. Post and F. L. Ribe, "Fusion Reactors as Future Energy Sources," *Science*, vol. 186, 1974, pp. 397–407.

94. Will nuclear energy be cheap? See A. L. Hammond, "The Fast Breeder Reactor: Signs of a Critical Reaction," *Science*, vol. 176, 1972, pp. 391–93, See also T. O'Toole, "Breeder Reactor Cost Soars," *Washington Post*, Feb. 5, 1975, p. E-16.

HOTHOUSE EARTH

94. Can we look forward to hundreds and thousands of nuclear reactors poised on the landscape and floating on the oceans? See *Business Week*, Feb. 9, 1974, pp. 57–58.
95. See A. M. Weinberg, "Social Institutions and Nuclear Energy," *Science*, vol. 177, 1972, pp. 27–34.

CHAPTER 7: POINT/COUNTERPOINT

98. The unattributed quotation was obtained in a phone conversation by John Fried.
100. The unattributed quotation was obtained in a private communication.
The unattributed quotation was obtained in a phone conversation by John Fried.
101. W. W. Kellogg, "Long Range Influences of Mankind on the Climate," National Center for Atmospheric Research, Boulder, Colo., 1974, p. 6.
103. To see that Dr. Kellogg is assuming in the paper referenced in regard to p. 101 that man will level off his rate of consumption of energy at about one per cent of the solar input rate, study pages 9 and 10 of that paper.
104. For data on the rapidity of ice cap response to global temperature changes, see, for example, W. D. Sellers, "A New Global Climatic Model," *Journal of Applied Meteorology*, vol. 12, 1973, p. 251.
See also J. O. Fletcher, "Polar Ice and the Global Climate Machine," *Bulletin of the Atomic Scientists*, Dec. 1970, p. 41.
A. Weinberg, "Global Effects of Man's Production of Energy," *Science*, vol. 186, 1974, p. 205.
106. For a summary statement regarding the current cooling trend for the earth, see W. W. Kellogg, "Long Range Influences of Mankind on the Climate," National Center for Atmospheric Research, Boulder, Colo., 1974, p. 5.
For signs of a coming surge of cold weather, see G. J. Kukla and R. K. Matthews, "When Will the Present Interglacial End?" *Science*, vol. 178, 1972, pp. 190–191. See also J. M. Mitchell, Jr., "On the World-wide Pattern of Secular Temperature Change," *Changes of Climate, Arid Zone Research XX*, UNESCO, Paris, 1963, pp. 161–180. (However, see also the remarks by J. M. Mitchell, Jr., on page 27 of the article by A. Anderson, Jr., "Forecast for Forecasting: Cloudy," *New York Times Magazine*, Dec. 29, 1974, p. 10.)

NOTES AND REFERENCES

107. See the paper by Kukla and Matthews cited in reference for page 106.

108. W. W. Kellogg, "Long-Range Influences of Mankind on the Climate," National Center for Atmospheric Research, Boulder, Colo., 1974, p. 10.

109. For a good discussion of many of the ways man is influencing or can influence the global climate, see W. H. Matthews, W. W. Kellogg, and G. D. Robinson, eds., *Man's Impact on Climate*, MIT Press, Cambridge, Mass., 1971.

110. For urban-rural temperature differences, see J. T. Peterson, "The Climate of Cities," U.S. Government Printing Office, Washington, 1969.
For data concerning the annual energy levels put out by various cities and regions, see *Report of the Study of Man's Impact on Climate*, MIT Press, Cambridge, Mass., 1971, p. 58.

112. For a discussion of the energy system designed for the World Trade Center in New York City, see "World Trade Center," *Power*, Jan. 1970, pp. 149–164.

113. For a discussion of the promising systems called "heat pumps," see T. Baumeister, "Heat Pump," *Encyclopedia of Science and Technology*, McGraw-Hill, New York, 1971, vol. 6, pp. 417–419.

114. A "nucleating agent" is needed for starting the formation of cloud droplets, raindrops, or snowflakes. Such an agent can be a natural dust particle, for example, or a tiny, man-made crystal. See B. J. Mason, "Cloud Physics," *Encyclopedia of Science and Technology*, McGraw-Hill, New York, 1971, vol. 3, pp. 226–230.

116. H. Flohn, *Climate and Weather*, McGraw-Hill, New York, 1969, p. 243.

CHAPTER 8:   SOLAR ENERGY AND OPEN-OCEAN FARMS

118. See H. A. Wilcox, "The Ocean Food and Energy Farm Concept," a paper presented to the Annual Meeting of the American Association for the Advancement of Science in New York City on Jan. 29, 1975, available by writing to Code 0103, Naval Undersea Center, San Diego, Calif., 92132.

119. See W. J. North and H. A. Wilcox, "History, Present Status, and Future Prospects Regarding the Experimental 7-Acre Marine Farm at San Clemente Island," Jan. 31, 1975, available by writing to Code 0103, Naval Undersea Center, San Diego, Calif. 92132.

PAGE

120. For information concerning the development of synthetic fuels and other products from the seaweeds harvested from ocean farms, see T. M. Leese, "Ocean Food & Energy Farm Kelp Product Conversion," a paper presented to the Annual Meeting of the American Association for the Advancement of Science in New York City on Jan. 29, 1975, available by writing to T. M. Leese, Code 402, Naval Weapons Center, China Lake, Calif. 93555, or to Code 0103, Naval Undersea Center, San Diego, Calif. 92132. Regarding solar-energy-gathering satellites, see P. E. Glaser, "Solar Power Via Satellite," Hearing Before the Committee on Aeronautical and Space Sciences, U.S. Senate, Oct. 31, 1973, pp. 11–62.

121. For the quotations given, see United States Energy: A Summary Review, U.S. Dept. of Interior, Jan. 1972, p. 34.

122. For a good discussion of solar cells and their potential, see A. L. Hammond, "Photovoltaic Cells: Direct Conversion of Solar Energy," Science, vol. 178, 1972, pp. 732–733.

123. The "inherent" expensiveness of sending payloads out into space results from (1) the relatively enormous energy expenditures required to inject them into orbit, and (2) the difficulties standing in the way of maintaining and repairing them (or else making them so long-lived and reliable that they do not need maintenance and repair) over many years of operation in orbit.

125. For a history of energy sources and energy usage in the USA, see P. C. Putnam, Energy in the Future, Van Nostrand, New York, 1953.
   For a discussion of wind-powered energy converters, see P. C. Putnam, Energy in the Future, Van Nostrand, New York, 1953, pp. 188–191.

126. For a discussion of turbines to be powered by ocean currents, see W. E. Heronemus, A Research Proposal Submitted to the National Science Foundation to Investigate a National Network of Pollution-Free Energy Sources, University of Massachusetts, Amherst, 1971, Appendix 3.
   For a discussion of energy systems powered by oceanic temperature differences, see A. Lavi and C. Zener, "Plumbing the Ocean Depths: A New Source of Power," IEEE Spectrum, Oct. 1973, pp. 22–27.

127. For a more complete discussion, with references, of the ocean farm system, see the paper cited in the note to page 118.
   See Govindjee, "Photosynthesis," Encyclopedia of Science and Technology, McGraw-Hill, New York, 1971, vol. 10, pp. 201–210.

PAGE

129. Regarding dwindling oceanic harvests, see L. Brown, *By Bread Alone*, Praeger, New York, 1974, pp. 148–150.

130. For the current biological productivity of the ocean and an estimate of its future potential, see J. E. Bardach, *Harvest of the Sea*, Harper & Row, New York, 1968, pp. 178–181.

131. For fish harvesting techniques, see J. E. Bardach, *Harvest of the Sea*, Harper & Row, New York, 1968, pp. 120–137.

132. For the eating of and other uses of seaweeds by ancient peoples, the Japanese, the Americans, et al., see C. P. Idyll, "The Harvest of Seaweed," Chapter 5 of *The Sea Against Hunger*, T. Y. Crowell, New York, 1970, pp. 47–63.

133. For seaweed farming and processing details, see the papers cited for pages 118 and 120.

134. Regarding the listed uses of seaweeds, see the pamphlet *Kelp*, Kelco Co., 2145 E. Belt St., San Diego, Calif. 92113.

135. For ability of *Macrocystis* to convert solar energy at high efficiency, see K. H. Mann, "Seaweeds: Their Productivity and Strategy for Growth," *Science*, vol. 182, 1973, pp. 975–981.

138. Regarding the native habitats and biological productivity of the kelps, see K. H. Mann, "Seaweeds: Their Productivity and Strategy for Growth," *Science*, vol. 182, 1973, fig. 1, p. 976.
    In reference to the fact that 99.9 percent of the ocean's surface waters are so barren of nutrients that they produce relatively little of the total life of the sea, see D. F. Othmer and O. A. Roels, "Power, Fresh Water, and Food from Cold, Deep Sea Water," *Science*, vol. 182, 1973, pp. 121–125.

139. For a discussion of the nutrients present in the waters of the ocean, see L. M. Jeffrey, "Nutrients in the Sea," in Fairbridge, ed., *Encyclopedia of Oceanography*, Reinhold, New York, 1966, pp. 557–562.

140. For a discussion of the power levels required and available for upwelling deep ocean water, see H. A. Wilcox, "Artificial Oceanic Upwelling," available by writing to Code 0103, Naval Undersea Center, San Diego, Calif. 92132.

141. For a discussion of Japanese experience with growing and eating seaweeds and other organisms, see W. N. Shaw, ed., "Proceedings of the First U.S.-Japan Meeting on Aquaculture," U.S. Government Printing Office, Washington, 1974.
    In reference to the possibility that Asians have special enzymes that enable them to digest seaweeds more effectively than Occidentals can, see C. P. Idyll, "The Harvest of Seaweed," Chapter

179

PAGE

5 of *The Sea Against Hunger,* T. Y. Crowell, New York, 1970, p. 54.

146. To check the numbers given in the text, note that the earth's total surface area is some 197 million square miles; that about 71 percent of this total—some 140 million square miles—is ocean; and roughly half of that ocean area lies between 30° north latitude and 30° south latitude. The solar energy absorbed at the surface of the earth is roughly 2.5×10^{21} Btu per year, so the world-wide average solar energy per unit area is about 1.3×10^{13} Btu per square mile per year. If 2 percent of that is converted to stored vegetational energy on each acre of ocean farm, the vegetational energy so produced will represent 2.6×10^{11} Btu per square mile per year. In 1975, man's world-wide consumption of petroleum, coal, and other such materials was about 2×10^{17} Btu, and his total population was about 4×10^{9} persons, so his world-wide consumption of energy was about 5×10^{7} Btu per person. If the farm system's conversion of vegetational energy to petroleum-like products can be carried out at 50 percent efficiency, then each square mile of farm will yield some 1.3×10^{11} Btu per year of such products, enough to support some 2600 persons at 1975 world average levels of consumption. Hence 78 million square miles of ocean farms could be expected to support some 200 billion persons at 1975 world average levels of consumption.

The last paragraph of Chapter 8 suggests the important question: What is the optimum population of humans on earth? The evidence and argumentation presented in this book are clearly relevant (though not sufficient) for finding the answer. If one assumes that man's optimum population is related to his achieving the greatest possible ability to cope with challenges to his survival as a species, then clearly he needs a large enough population on earth to be able to compensate for variations in the rate of solar energy input to the planet, variations that could easily cause ice ages, surges of sea level, or other major climatic upsets. This leads to the idea that he should be able to control—either to increase or to decrease—the earth's reflectivity and surface energy dissipation rate by amounts ranging up to 10 percent of their present values. If one next assumes that man's per capita influence on these variables will continue to grow at 3 percent per year, say, and that his population will continue also to grow at some 2 percent per year, then man can reach the indicated level of global climatic control by about the year 2100 A.D. at which time his global population would be about 50 billion individuals.

CHAPTER 9:   EPILOGUE

PAGE

150. *New York Times,* Aug. 17, 1972, p. 35.
151. That "our leaders seem mesmerized by nuclear power and . . .
seem determined to make it the main source of energy to meet
our future needs" is attested to by President Gerald R. Ford's
1975 State of the Union Message, in which he said: "Within
the next 10 years, my program envisions 200 major nuclear power
plants . . ."

The Author:

DR. HOWARD A. WILCOX has devoted thirty-two years of professional life to the questions discussed in *Hothouse Earth*. Educated as a physicist at the Universities of Minnesota, Harvard, and Chicago, he spent the years during and immediately after World War II at Los Alamos, New Mexico, the Fermi Institute of Nuclear Studies at Chicago, and the Lawrence Radiation Laboratory at Berkeley, doing basic and applied research in nuclear physics and weapons. In late 1950 he became project engineer, then manager, of the SIDEWINDER heat-seeking guided missile development project. After several years as deputy director of the U.S. Defense Department's world-wide research and engineering program, Wilcox became director of research and engineering for the Research Laboratories of General Motors Corporation in Santa Barbara, California. In 1967 he became chairman of that city's Environmental Quality Advisory Board, and in 1974 he moved to the Naval Undersea Center in San Diego, California, to manage the Ocean Farm Project.

In regard to his activity in such seemingly unrelated fields as marine biology, climatology, and the dynamics of large social systems, Wilcox believes that "the physicist is always the generalist among experts, and the prime skill required of the manager of any development project is to know what can be accomplished without knowing how to accomplish it."

Dr. Wilcox has published some fifty original contributions in scientific and semitechnical publications. This is his first book.

JOHN J. FRIED, who collaborated with Dr. Wilcox in the writing of this book, is a free-lance writer who specializes in scientific and medical affairs. His books include *The Mystery of Heredity* (1971), *Vasectomy—Truth and Consequences* (1972), *Life Along the San Andreas Fault* (1973), *Code Arrest: A Heart Stops* (with Edward B. Diethrich, M.D.) (1974), and *The Vitamin Conspiracy* (1975). Mr. Fried's articles have appeared in the *New York Times Magazine*, *Sports Illustrated*, *Playgirl*, *Reader's Digest*, *Family Health*, and *Esquire*.